I0488103

Simulated Effects of Hydrologic, Water Quality, and Land-Use Changes of the Lake Maumelle Watershed, Arkansas, 2004–10

By Rheannon M. Hart, W. Reed Green, Drew A. Westerman, James C. Petersen, and Jeanne L. De Lanois

Prepared in cooperation with Central Arkansas Water

Scientific Investigations Report 2012–5246
(Revised February 2013)

U.S. Department of the Interior
U.S. Geological Survey

U.S. Department of the Interior
KEN SALAZAR, Secretary

U.S. Geological Survey
Marcia K. McNutt, Director

U.S. Geological Survey, Reston, Virginia: 2012
(Revised 2013)

This and other USGS information products are available at http://store.usgs.gov/

U.S. Geological Survey
Box 25286, Denver Federal Center
Denver, CO 80225

To learn about the USGS and its information products visit http://www.usgs.gov/
1-888-ASK-USGS

Suggested citation:
Hart, R.M., Green, W.R., Westerman, D.A., Petersen, J.C., and De Lanois, J.L., 2012, Simulated effects of hydrologic, water quality, and land-use changes of the Lake Maumelle watershed, Arkansas, 2004–10: U.S. Geological Survey Scientific Investigations Report 2012–5246, 119 p. (Revised February 2013)

Contents

Figures

Tables

Conversion Factors

Inch/Pound to SI

Multiply	By	To obtain
Length		
inch (in.)	2.54	centimeter (cm)
foot (ft)	0.3048	meter (m)
mile (mi)	1.609	kilometer (km)
yard (yd)	0.9144	meter (m)
Area		
acre	4,047	square meter (m²)
square mile (mi²)	2.590	square kilometer (km²)
Volume		
acre-foot (acre-ft)	1,233	cubic meter (m³)
Flow rate		
cubic foot per second (ft³/s)	0.02832	cubic meter per second (m³/s)
inch per hour (in/h)	0 .0254	meter per hour (m/h)
Mass		
pound per day (lb/d)	453.6	gram per day (g/d)
ton per day (ton/d)	0.9072	metric ton per day
ton per day (ton/d)	0.9072	megagram per day (Mg/d)
Application rate		
pound per acre per year [(lb/acre)/yr]	1.121	kilograms per hectare per year [(kg/ha)/yr]
tons per acre per year [(ton/acre)/yr]	2,242	kilograms per hectare per year [(kg/ha)/yr]
cubic foot per second (ft³/s)	0.02832	cubic meter per second (m³/s)

Temperature in degrees Celsius (°C) may be converted to degrees Fahrenheit (°F) as follows:

$$°F=(1.8×°C)+32$$

Temperature in degrees Fahrenheit (°F) may be converted to degrees Celsius (°C) as follows:

$$°C=(°F-32)/1.8$$

Vertical coordinate information is referenced to the National Geodetic Vertical Datum of 1929 (NGVD 1929).

Horizontal coordinate information is referenced to the North American Datum of 1983 (NAD 1983).

Altitude, as used in this report, refers to distance above the vertical datum.

Concentrations of chemical constituents in water are given either in milligrams per liter (mg/L) or micrograms per liter (µg/L).

Water year (WY)—Is the period October 1 through September 30 designated by the calendar year in which it ends, for example, the 2008 water year runs from October 1, 2007 to September 30, 2008 (Rantz and others,1982).

Acronyms and Abbreviations

λ – light extinction coefficient

ACQOP – accumulation rate of the constituent

AMLE – adjusted maximum likelihood estimation

ARM – Agricultural Runoff Model

CAW – Central Arkansas Water

CBHE – coefficient of bottom heat exchange

FTABLE – function table

GIS – geographic information system

GQUAL – generalized quality constituent

HRAP – Hydrologic Rainfall Analysis Project

HSPF – Hydrologic Simulation Program–FORTRAN (HSPF) watershed model

IMPLND – impervious land

INFILT – index to the infiltration capacity of the soil

KATRAD – atmospheric longwave radiation coefficient

LDOM – labile dissolved organic matter

LRL – laboratory reporting level

LSUR – length of the assumed overland flow plane

LT–MDL – long-term method detection level

LZETP – lower zone evapotranspiration parameter

LZSN – lower zone nominal storage

MAE – mean absolute error

MLE – maximum likelihood estimation

MPE – Multisensor Precipitation Estimator

NCDC – National Climatic Data Center

NED – National Elevation Dataset

NEXRAD – Next-Generation Radar; a Doppler weather radar system used to obtain precipitation and wind data

NHD – National Hydrography Dataset

NPS – Nonpoint Source

NSE – Nash-Sutcliffe model efficiency coefficient

NSUR – Manning's n for the overland flow plane

PERLND – pervious land

POTFW – washoff potency factor

RCHRES – reach reservoir

RETSC – retention (interception) storage capacity of the surface

RFCs – River Forecasting Centers

RMSE – root mean square error

RPD – relative percentage difference

R2 – coefficient of determination

SEP – standard error of prediction

SLSUR – slope of the overland flow plane

SPARROW – SPAtially-Referenced Regression on Watershed attributes

SQOLIM – asymptotic limit for storage of available quality constituent of the surface as time approaches infinity

SSURGO – Soil Survey Geographic Database

USGS – U.S. Geological Survey

UWA – Upper Watershed Area

UZSN – upper zone nominal storage

WSC – wind-sheltering coefficient

WSQOP – rate of surface runoff that will remove 90 percent of stored quality constituent in 1 hour

WSR-88D – Weather Surveillance Radar – 1998 Doppler

Simulated Effects of Hydrologic, Water Quality, and Land-Use Changes of the Lake Maumelle Watershed, Arkansas, 2004–10

By Rheannon M. Hart, W. Reed Green, Drew A. Westerman, James C. Petersen, and Jeanne L. De Lanois

Abstract

Lake Maumelle, located in central Arkansas northwest of the cities of Little Rock and North Little Rock, is one of two principal drinking-water supplies for the Little Rock, and North Little Rock, Arkansas, metropolitan areas. Lake Maumelle and the Maumelle River (its primary tributary) are more pristine than most other reservoirs and streams in the region with 80 percent of the land area in the entire watershed being forested. However, as the Lake Maumelle watershed becomes increasingly more urbanized and timber harvesting becomes more extensive, concerns about the sustainability of the quality of the water supply also have increased.

Two hydrodynamic and water-quality models were developed to examine the hydrology and water quality in the Lake Maumelle watershed and changes that might occur as the watershed becomes more urbanized and timber harvesting becomes more extensive. A Hydrologic Simulation Program–FORTRAN watershed model was developed using continuous streamflow and discreet suspended-sediment and water-quality data collected from January 2004 through 2010. A CE–QUAL–W2 model was developed to simulate reservoir hydrodynamics and selected water-quality characteristics using the simulated output from the Hydrologic Simulation Program–FORTRAN model from January 2004 through 2010.

The calibrated Hydrologic Simulation Program–FORTRAN model and the calibrated CE–QUAL–W2 model were developed to simulate three land-use scenarios and to examine the potential effects of these land-use changes, as defined in the model, on the water quality of Lake Maumelle during the 2004 through 2010 simulation period. These scenarios included a scenario that simulated conversion of most land in the watershed to forest (scenario 1), a scenario that simulated conversion of potentially developable land to low-intensity urban land use in part of the watershed (scenario 2), and a scenario that simulated timber harvest in part of the watershed (scenario 3). Simulated land-use changes for scenarios 1 and 3 resulted in little (generally less than

10 percent) overall effect on the simulated water quality in the Hydrologic Simulation Program–FORTRAN model. The land-use change of scenario 2 affected subwatersheds that include Bringle, Reece, and Yount Creek tributaries and most other subwatersheds that drain into the northern side of Lake Maumelle; large percent increases in loading rates (generally between 10 and 25 percent) included dissolved nitrite plus nitrate nitrogen, dissolved orthophosphate, total phosphorus, suspended sediment, dissolved ammonia nitrogen, total organic carbon, and fecal coliform bacteria.

For scenario 1, the simulated changes in nutrient, suspended sediment, and total organic carbon loads from the Hydrologic Simulation Program–FORTRAN model resulted in very slight (generally less than 10 percent) changes in simulated water quality for Lake Maumelle, relative to the baseline condition. Following lake mixing in the falls of 2006 and 2007, phosphorus and nitrogen concentrations were higher than the baseline condition and chlorophyll *a* responded accordingly. The increased nutrient and chlorophyll *a* concentrations in late October and into 2007 were enough to increase concentrations, on average, for the entire simulation period (2004–10). For scenario 2, the simulated changes in nutrient, suspended sediment, total organic carbon, and fecal coliform bacteria loads from the Lake Maumelle watershed resulted in slight changes in simulated water quality for Lake Maumelle, relative to the baseline condition (total nitrogen decreased by 0.01 milligram per liter; dissolved orthophosphate increased by 0.001 milligram per liter; chlorophyll *a* decreased by 0.1 microgram per liter). The differences in these concentrations are approximately an order of magnitude less than the error between measured and simulated concentrations in the baseline model. During the driest summer in the simulation period (2006), phosphorus and nitrogen concentrations were lower than the baseline condition and chlorophyll *a* concentrations decreased during the same summer season. The decrease in nitrogen and chlorophyll *a* concentrations during the dry summer in 2006 was enough to decrease concentrations of these constituents very slightly, on average, for the entire simulation period (2004–10). For scenario 3, the changes in simulated nutrient, suspended sediment, total organic carbon, and fecal coliform bacteria loads from Lake Maumelle watershed resulted in very slight changes in simulated water quality within Lake Maumelle, relative to the baseline condition, for most of the reservoir.

Among the implications of the results of the modeling described in this report are those related to scale in both space and time. Spatial scales include limited size and location of land-use changes, their effects on loading rates, and resultant effects on water quality of Lake Maumelle. Temporally, the magnitude of the water-quality changes simulated by the land-use change scenarios over the 7-year period (2004–10) are not necessarily indicative of the changes that could be expected to occur with similar land-use changes persisting over a 20-, 30-, or 40- year period, for example. These implications should be tempered by realization of the described model limitations.

The Hydrologic Simulation Program–FORTRAN watershed model was calibrated to streamflow and water-quality data from five streamflow-gaging stations, and in general, these stations characterize a range of subwatershed areas with varying land-use types. The CE–QUAL–W2 reservoir model was calibrated to water-quality data collected during January 2004 through December 2010 at three reservoir stations, representing the upper, middle, and lower sections of the reservoir.

In general, the baseline simulation for the Hydrologic Simulation Program–FORTRAN and the CE–QUAL–W2 models matched reasonably well to the measured data. Simulated and measured suspended-sediment concentrations during periods of base flow (streamflows not substantially influenced by runoff) agree reasonably well for Maumelle River at Williams Junction, the station representing the upper end of the watershed (with differences—simulated minus measured value—generally ranging from -15 to 41 milligrams per liter, and percent difference—relative to the measured value—ranging from -99 to 182 percent) and Maumelle River near Wye, the station just above the reservoir at the lower end (differences generally ranging from -20 to 22 milligrams per liter, and percent difference ranging from -100 to 194 percent). In general, water temperature and dissolved-oxygen concentration simulations followed measured seasonal trends for all stations with the largest differences occurring during periods of lowest temperatures or during the periods of lowest measured dissolved-oxygen concentrations.

For the CE–QUAL–W2 model, simulated vertical distributions of water temperatures and dissolved-oxygen concentrations agreed with measured vertical distributions over time, even for the most complex water-temperature profiles. Considering the oligotrophic-mesotrophic (low to intermediate primary productivity and associated low nutrient concentrations) condition of Lake Maumelle, simulated algae, phosphorus, and nitrogen concentrations compared well with generally low measured concentrations.

Introduction

Lake Maumelle, located in central Arkansas northwest of the city of Little Rock, is one of two principal drinking-water supplies for the Little Rock, and North Little Rock, Arkansas, metropolitan areas. As the Lake Maumelle watershed becomes increasingly more urbanized and timber harvesting becomes more extensive, concerns about the sustainability of the quality of the water supply also have increased as documented in Tetra Tech, Inc. (2007). Using streamflow and water-quality data collected by the U.S. Geological Survey (USGS) in cooperation with Central Arkansas Water (CAW), watershed and reservoir models previously were developed (Green, 2001; Tetra Tech, Inc., 2006) to evaluate water-quality characteristics of Lake Maumelle. However, evaluation is needed of possible effects of anthropogenic (human-induced) activities in the watershed on the drinking-water supply, and additional data have been collected since the previous models were developed. Therefore, the USGS, in cooperation with CAW, has enhanced the previous models by refining the Lake Maumelle reservoir grid using a digitized preimpoundment map of elevation contours, as well as using more detailed precipitation (Next Generation Radar), streamflow, and water-quality data collected through 2010 for calibration and has applied the models to address different potential land-use changes within the watershed.

Purpose and Scope

The purpose of this report is to present the findings of developed watershed and reservoir models used to simulate the effects of hydrologic, water quality, and land-use changes of the Lake Maumelle watershed. The development and results of a coupled Hydrologic Simulation Program–FORTRAN (HSPF) (Bicknell and others, 2001) watershed model (hereafter referred to as the HSPF model) of the Lake Maumelle watershed and a two-dimensional CE–QUAL–W2 (Cole and Wells, 2008) reservoir model (hereafter referred to as the CE–QUAL–W2 model) of Lake Maumelle are described. The HSPF model was developed to simulate streamflow, water temperature, dissolved oxygen, suspended sediment, total organic carbon, dissolved ammonia nitrogen, dissolved nitrite plus nitrate nitrogen, dissolved orthophosphate, total phosphorus, and fecal coliform bacteria using input data collected from January 2004 through December 2010. The CE–QUAL–W2 model was developed to simulate reservoir hydrodynamics and selected water-quality characteristics including water temperature, dissolved-oxygen concentrations, nutrient concentrations, organic-carbon concentrations, algae, and chlorophyll *a* concentrations using the simulated output from the HSPF model from January 2004 through December 2010. The HSPF and CE–QUAL–W2 models were used to simulate three scenarios, which represent a range of potential generalized land-use changes. The

HSPF and CE–QUAL–W2 model input data are described, followed by a discussion of the development, calibration, and application of the models.

Description of Study Area

Lake Maumelle is located in central Arkansas northwest of the city of Little Rock (fig. 1). Dam construction was completed in 1958 (Martin Maner, Central Arkansas Water, written commun., 2010). The lake contains approximately 219,000 acre-feet (acre-ft) when the water surface is at the spillway altitude (290 feet above National Geodetic Vertical Datum (NGVD) of 1929) (Green, 2001). The surface area of the pool at the spillway altitude is approximately 14 square miles (mi^2). The maximum length of the reservoir is 12 miles (mi), and the maximum depth is 46 feet (ft) with a mean depth of 25 ft. The drainage area upstream from the spillway is 137 mi^2 (Green, 2001).

The Lake Maumelle watershed lies in the Ouachita Mountains physiographic section (Fenneman and Johnson, 1946; fig. 1). The Ouachita Mountains are an east-west trending mountain range consisting primarily of sandstone, shale, novaculite, and chert. The land-surface altitude within the Lake Maumelle watershed ranges from approximately 1,500 ft at a pinnacle at the western edge of the watershed to 290 ft at the Lake Maumelle Spillway (U.S. Geological Survey, 1994).

The major soils in the watershed consist primarily of the Carnasaw series and are described as well drained, gently sloping to steep, moderately deep and shallow, loamy and stony soils (U.S. Department of Agriculture, 1975a, 1975b, 1975c). Most of the soils within the watershed are characterized by low infiltration rates when thoroughly wetted.

Approximately 80 percent of the land area within the Lake Maumelle watershed is forest, approximately 10 percent water (including Lake Maumelle), approximately 5.6 percent clearcut area, and approximately 3 percent grasslands (Arkansas Natural Resources Commission and University of Arkansas: Center for Advanced Spatial Technologies, 2009; Martin Maner, Central Arkansas Water, written commun., 2010). The remaining approximate 1.4 percent is divided between agriculture, bare soil, urban, and paved roads (fig. 2). Recent aerial photography data (2009) indicate that there are approximately 549 mi of roads in the watershed and approximately 474 mi are unpaved (Martin Maner, Central Arkansas Water, written commun., 2010). No point-source dischargers exist in the watershed (Arkansas Department of Environmental Quality, 2008). Future land-use changes are expected in the watershed; up to 53 percent of the watershed is potentially developable (Tetra Tech, Inc., 2007).

Lake Maumelle is a water-supply reservoir for the Little Rock and North Little Rock metropolitan areas including 15 cities and communities in central Arkansas serving approximately 388,000 people in 2007 (Tetra Tech, Inc., 2007). In addition to water supply, the reservoir is used for recreation and fish and wildlife habitat. Changes in land use and potential changes in water quality are a concern as documented in Tetra Tech, Inc. (2007). Monitoring changes in the hydrology and water quality of the watershed as the land-use changes is critical to managing the resource.

As part of CAW's plans for managing Lake Maumelle, a watershed management plan has been developed by Tetra Tech, Inc., CAW, stakeholders, and State and local resource agencies and institutions (Tetra Tech, Inc., 2007). One of the primary goals of the plan is to maintain Lake Maumelle as a high quality drinking-water supply. Tetra Tech, Inc., with review and comment from an advisory group, recommended water-quality targets (for chlorophyll *a*, total organic carbon, turbidity or Secchi disk, and fecal coliform bacteria) and associated numeric targets (performance standards for total phosphorus, total suspended solids, and total organic carbon loading rates for three management areas) for assessing compliance with the goals and objectives. Three management areas were defined: Critical Area A, closest to the intake, would have the most restrictive proposed requirements (including the performance standards); Critical Area B, surrounding the lake but with a longer traveltime to the lake and intake than Critical Area A, would have less stringent requirements than Critical Area A; the Upper Watershed Area (UWA) would have the least restrictive requirements (fig. 1).

Previous Investigations

The USGS has collected reservoir pool altitude, streamflow, and water-quality data from Lake Maumelle and the main tributary, the Maumelle River, since May 1989 as part of an ongoing monitoring program in cooperation with CAW, the utility that owns and operates the water supply reservoir (Galloway and Green, 2004) (tables 1 and 2).

A long-term water-quality database has been developed by monitoring the hydrology and water quality of Lake Maumelle and associated tributaries. These data are stored in the USGS National Water Information System database and published annually (Morris and others, 1992; Westerfield and others, 1994; Evans and others, 1995; Porter and others, 1993, 1996, 1997, 1998, 1999, 2000, 2001, 2002; Brossett and Evans, 2003; Brossett and others, 2005; Evans and others, 2004; Schrader and others, 2006; U.S. Geological Survey, 2007, 2008, 2009a, 2010). Hydrologic data collected for Lake Maumelle from May 1989 to October 1992 (Green and Louthian, 1993) were used to assess the water quality of the lake, and Green (1994) concluded that Lake Maumelle and the Maumelle River are more pristine than most other reservoirs and streams in the region. In a later report, Green (2001) concluded that nutrient concentrations in Lake Maumelle and the Maumelle River were one to two orders of magnitude lower than estimates of national background nutrient concentrations in streams.

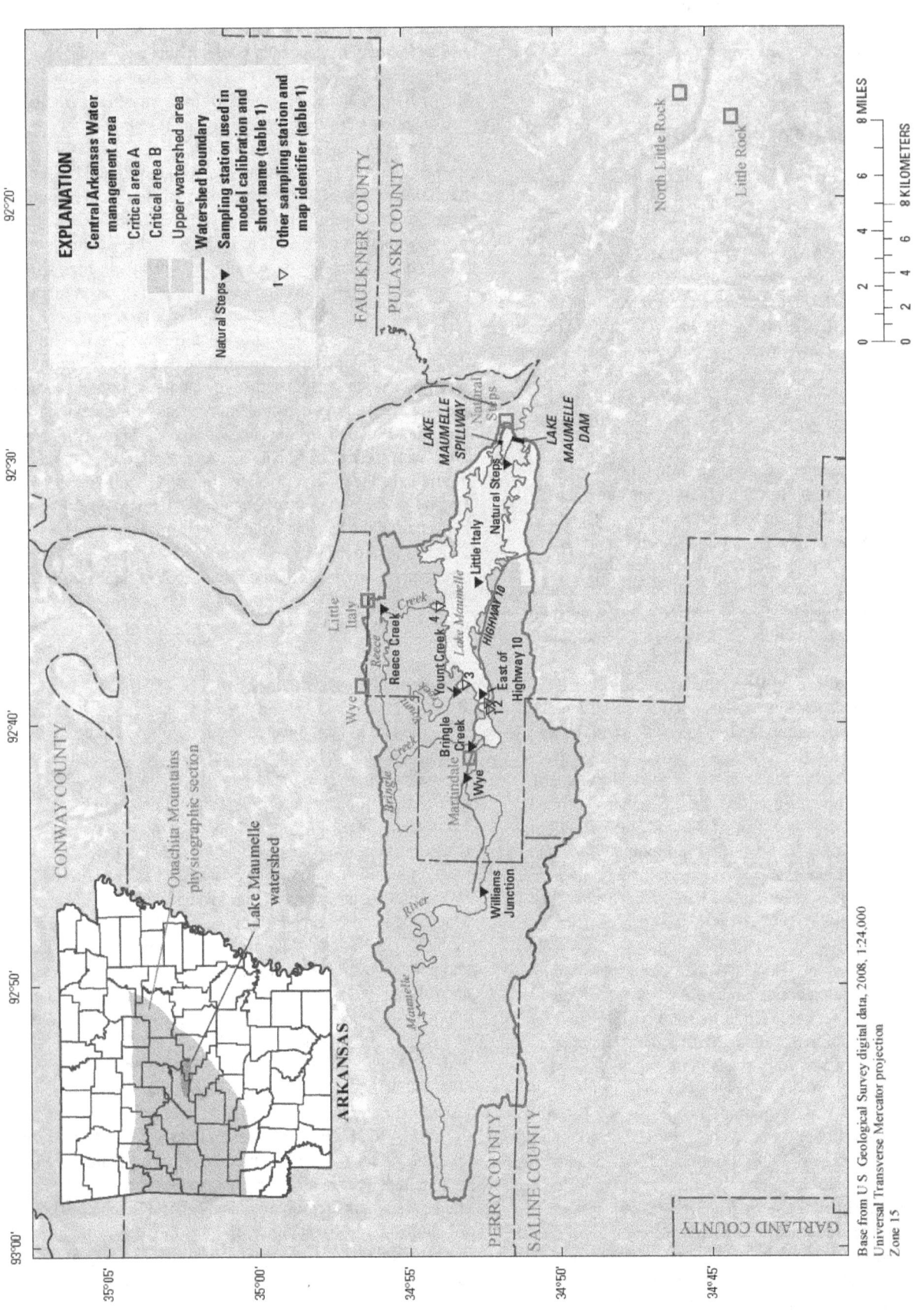

Figure 1. Location of study area, Lake Maumelle, Arkansas.

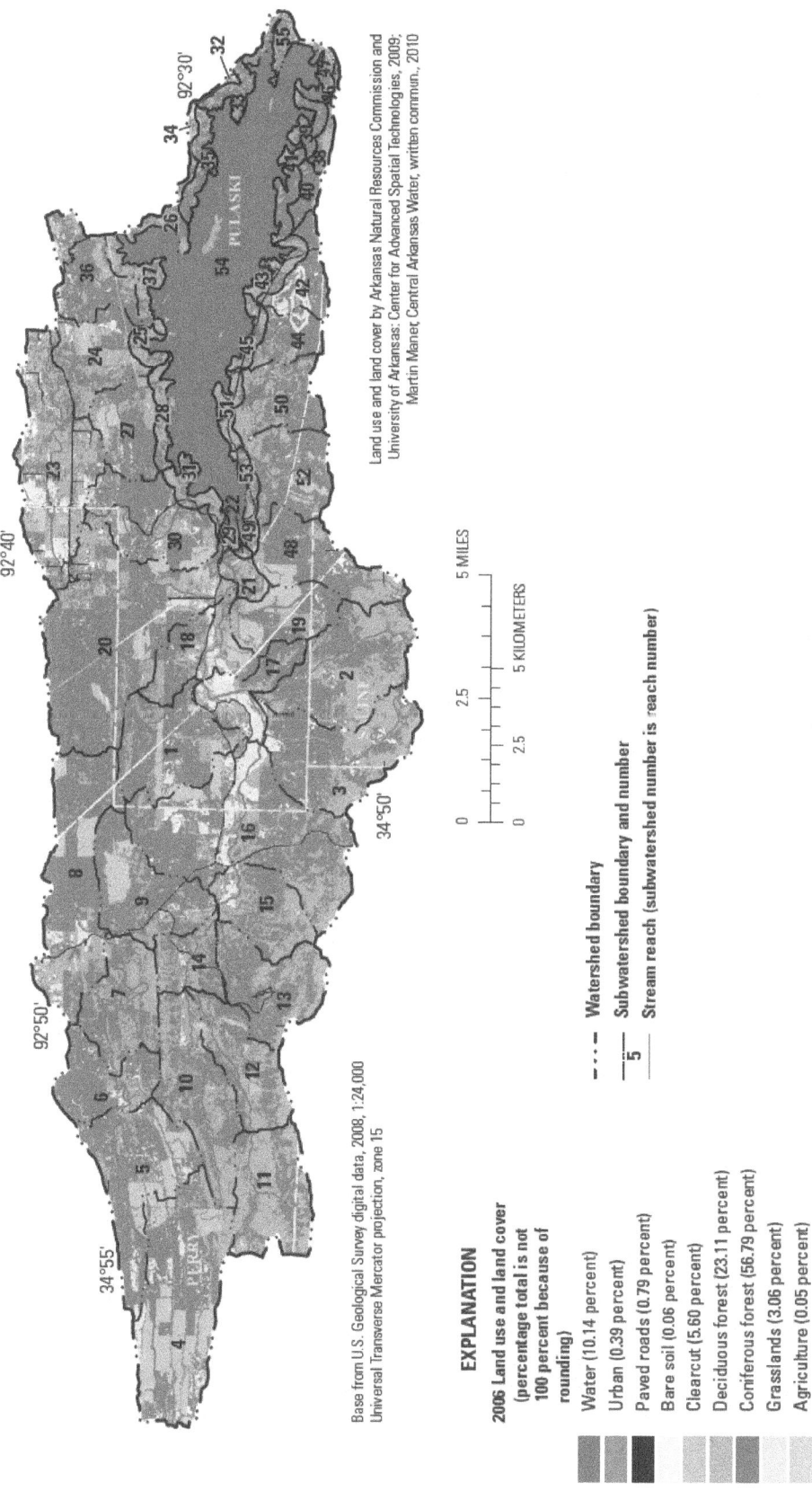

Base from U.S. Geological Survey digital data, 2008, 1:24,000
Universal Transverse Mercator projection, zone 15

Land use and land cover by Arkansas Natural Resources Commission and
University of Arkansas: Center for Advanced Spatial Technologies, 2009;
Martin Maner, Central Arkansas Water, written commun., 2010

EXPLANATION

2006 Land use and land cover
(percentage total is not
100 percent because of
rounding)

Water (10.14 percent)
Urban (0.39 percent)
Paved roads (0.79 percent)
Bare soil (0.06 percent)
Clearcut (5.60 percent)
Deciduous forest (23.11 percent)
Coniferous forest (56.79 percent)
Grasslands (3.06 percent)
Agriculture (0.05 percent)

– – – –   Watershed boundary

——   Subwatershed boundary and number
5

——   Stream reach (subwatershed number is reach number)

Figure 2.   Land-use types within the study area, Lake Maumelle, Arkansas.

Table 1. Station information for water-quality and streamflow monitoring stations in the Lake Maumelle watershed, Arkansas.

[ddmmss, degrees, minutes, seconds; mi², square mile; Flow, streamflow data collected, QW, water-quality data collected; n/a, not applicable]

Station (USGS station identification number)	Short name or map identifier (fig. 1)	Station type	Period of record used for analysis	Latitude (ddmmss)	Longitude (ddmmss)	Drainage area (mi²)
Maumelle River at Williams Junction (07263295)[1,2]	Williams Junction	Flow, QW	October 1989–September 2011	345234	924628	46.1
Maumelle River near Wye (07263296)[1,2]	Wye	Flow, QW	June 2007–September 2011	345244	924108	73.0
Bringle Creek at Martindale (072632962)[2]	Bringle Creek	Flow, QW	May 2005–September 2011	345253	924052	8.70
Yount Creek near Martindale (072632971)[2]	Yount Creek	Flow, QW	May 2005–September 2011	345323	923848	2.45
Reece Creek at Little Italy (072632982)[2]	Reece Creek	Flow, QW	May 2005–September 2011	345547	923536	4.96

Station (station identification number)	Short name or map identifier (fig. 1)	Station type	Period of record used for analysis	Latitude (ddmmss)	Longitude (ddmmss)	Drainage area (mi²)
Lake Maumelle west of Highway 10 Bridge (072632965)[1]	1	QW	July 1991–September 1992, February 2000–September 2011	345224	923926	n/a
Lake Maumelle at Highway 10 Bridge (072632966)[2]	2	Flow, QW	October 2002–September 2011	345230	923913	n/a
Lake Maumelle east of Highway 10 Bridge (07263297)[1]	East of Highway 10	QW	May 1989–September 2011	345231	923853	n/a
Lake Maumelle at mouth of Yount Creek near Martindale (0726329710)[1]	3	QW	March 2006–September 2011	345315	923835	n/a
Lake Maumelle near Little Italy (07263299)[1]	Little Italy	QW	May 1989–September 2011	345234	923435	n/a
Lake Maumelle at mouth of Reece Creek near Martindale (072632990)[1]	4	QW	March 2006–September 2011	345401	923532	n/a
Lake Maumelle at Natural Steps (072632995)[1]	Natural Steps	QW	May 1989–September 2011	345139	923007	n/a

[1]Sampled monthly or quarterly and during selected storm events.

[2]Sampled during storm events only.

Table 2. Medians of constituent values from water samples collected at water-quality sampling stations in Lake Maumelle watershed, Arkansas, and its inflow stations (1989–2010 water years).

[A water year is the 12-month period October 1 through September 30 designated by the calendar year in which it ends; N, nitrogen; P, phosphorus; C, carbon; mg/L, milligrams per liter; col/100 mL, colonies per 100 milliliters; µg/L, micrograms per liter; <, less than; ---, no value; (00010), parameter code]

Station name (station identification number)	Dissolved ammonia (mg/L as N) (00608)	Dissolved nitrite plus nitrate (mg/L as N) (00631)	Dissolved orthophosphorus (mg/L as P) (00671)	Total phosphorus (mg/L as P) (00665)	Dissolved organic carbon (mg/L as C) (00681)
Median of water-quality characteristics for period of record[1]					
Inflow stations					
Maumelle River at Williams Junction (07263295)[2,3]	<0.02	0.026	<0.006	0.018	3.4
Maumelle River near Wye (07263296)[2,3]	<0.02	0.036	<0.006	0.021	3.6
Bringle Creek at Martindale (072632962)[3]	<0.02	0.066	<0.006	0.038	4.4
Yount Creek near Martindale (072632971)[3]	0.03	0.090	<0.006	0.035	6.1
Reece Creek at Little Italy (072632982)[3]	<0.02	0.060	<0.006	0.058	5.6
Lake stations					
Lake Maumelle West of Highway 10 Bridge (072632965)[2]	<0.02	<0.016	<0.006	0.018	3.2
Lake Maumelle East of Highway 10 Bridge (07263297)[2]	<0.02	<0.016	<0.006	0.016	3.3
Lake Maumelle at mouth of Yount Creek near Martindale (0726329710)[2]	<0.02	<0.016	<0.006	0.016	3.6
Lake Maumelle near Little Italy (07263299)[2]	<0.02	<0.016	<0.006	0.012	3.2
Lake Maumelle at mouth of Reece Creek near Martindale (072632990)[2]	<0.02	<0.016	<0.006	0.014	3.5
Lake Maumelle near Natural Steps (072632995)[2]	<0.02	<0.016	<0.006	0.011	3.1

Station name (station identification number)	Total organic carbon (mg/L as C) (00680)	Fecal coliform (col/100 mL) (31625)	Suspended sediment (mg/L) (80154)	Chlorophyll *a* (µg/L) (70953)	Transparency, secchi (ft) (00078)
Median of water-quality characteristics for period of record[1]					
Inflow stations					
Maumelle River at Williams Junction (07263295)[2,3]	4.3	83	11	---	---
Maumelle River near Wye (07263296)[2,3]	4.7	77	9	---	---
Bringle Creek at Martindale (072632962)[3]	6.2	930	21	---	---
Yount Creek near Martindale (072632971)[3]	6.9	1,100	14	---	---
Reece Creek at Little Italy (072632982)[3]	8.9	2,450	55	---	---
Lake stations					
Lake Maumelle West of Highway 10 Bridge (072632965)[2]	3.9	8	---	3.8	3.5
Lake Maumelle East of Highway 10 Bridge (07263297)[2]	4.1	4	---	3.2	4.3
Lake Maumelle at mouth of Yount Creek near Martindale (0726329710)[2]	4.2	4	---	3.8	4.5
Lake Maumelle near Little Italy (07263299)[2]	3.8	<1	---	3.5	6.6
Lake Maumelle at mouth of Reece Creek near Martindale (072632990)[2]	3.9	4	---	4.0	5.4
Lake Maumelle near Natural Steps (072632995)[2]	3.8	<1	---	3.3	6.9

[1]All data, including censored data, were used to calculate medians.

[2]Sampled monthly or quarterly.

[3]Sampled during storm events.

Galloway and Green (2004) assessed the water quality of Lake Maumelle based on the water-quality record from 1991 through 2003. Annual and seasonal loads of nutrients, dissolved organic carbon, and suspended sediment were estimated for the main inflows into Lake Maumelle. Loads vary seasonally with the highest daily loads in the winter and fall and the lowest daily loads in the summer. Yields and flow-weighted mean concentrations of nutrients were calculated from the estimated annual loads; calculations indicated that nutrient concentrations for the Maumelle River were similar to selected undeveloped sites across the Nation.

Green (2001) developed and calibrated a hydrodynamic and water-quality model of Lake Maumelle to simulate the temperature, dissolved oxygen, nutrient, and algal biomass dynamics in the reservoir from 1991 to 1992. The model was used to evaluate reservoir response to a hypothetical spill of a conservative material (no decay or production) at the upper end of Lake Maumelle. In addition, model simulations of the algal response to increases of nitrogen and phosphorus loads demonstrated Lake Maumelle is limited in phosphorus (Green, 2001), meaning sustained algal productivity is dependent on continuous or pulsed phosphorus loading into the system, and productivity is ceased when loading decreases. A previous model was developed to simulate the release of water and associated constituents from a nearby fish nursery pond into a tributary on the south side of Lake Maumelle (Green, 1998). Simulation results showed elevated concentrations of some nutrients, organic carbon, iron, and manganese during simulated releases in 1991 through 1994 and 1996 (Green, 1998).

Pomes and others (1997, 1999) evaluated the sources of disinfection byproduct precursors in Lake Maumelle. Aquatic humic substances that generate potentially harmful disinfection byproducts and dissolved organic carbon concentrations were found to be low in Lake Maumelle and likely originated from deciduous leaf litter, twigs, and grass leachates.

The models presented here are enhancements of the HSPF and CE–QUAL–W2 models developed by Tetra Tech, Inc. (2006). The purposes of Tetra Tech, Inc.'s modeling efforts were to address the response of Lake Maumelle to future changes in land-use and management practices, to establish management goals, evaluate management options and risks, and make recommendations for water-quality program enhancement (Tetra Tech, Inc., 2006). The calibration period covered by Tetra Tech, Inc. was water years 1997 through 2004 (October 1996 through September 2004).

Description of Water-Quality Data of Lake Maumelle and Inflows

Water-quality data (data available from the USGS National Water Information System, http://waterdata.usgs.gov/nwis) for inflow stations in the Lake Maumelle watershed (table 1) and water-quality data for the Lake Maumelle stations were used to calculate median constituent values for each station for the period of record through September 2011 (table 2). The inflow stations on the Maumelle River at Williams Junction and near Wye were sampled quarterly or monthly (monthly after January 2008) and during storm events (at varying annual frequency) by equal-width-increment sampling methods. The inflow stations on Bringle Creek at Martindale, Yount Creek near Martindale, and Reece Creek at Little Italy were sampled only during storm events using automatic and equal-width-increment sampling methods. All inflow samples were collected and processed using USGS protocols described in Wilde and Radke (1998), Wilde and others (1998a, 1998b, 1998c, 1999a, and 1999b), and Myers and Wilde (1999). The lake stations were sampled at varying frequencies since 1989 (most lake stations were sampled quarterly beginning in January 2001 and monthly beginning in January 2007). Since January 2002, all lake station samples were collected using point sampling techniques for the epilimnion (approximately 3.28-ft (1 meter [m]) below surface) and hypolimnion (approximately 3.28-ft (1 m) above the bottom); prior to January 2002 samples were collected using a method that composited several depths in the epilimnion or hypolimnion. Additional information about sampling of lake (and inflow) stations can be found in Galloway and Green (2004). Median values of water-quality constituents for the inflow and lake stations are summarized below.

Median concentrations of dissolved ammonia (as nitrogen) at most inflow stations (some of which were sampled only during storm events) were less than 0.02 milligrams per liter (mg/L). Maximum storm inflow concentration (0.134 mg/L, maximum values not shown on table 2) occurred at Yount Creek near Martindale. Median concentrations of dissolved ammonia (as nitrogen) at all lake stations were consistently at or below the laboratory reporting level (LRL) (less than 0.02 mg/L).

Median concentrations of dissolved nitrite plus nitrate (as nitrogen) at inflow stations (some of which were sampled only during storm events) ranged from 0.026 to 0.090 mg/L. Maximum storm inflow concentration (0.757 mg/L) occurred at Maumelle River near Wye. Median concentrations of dissolved nitrite plus nitrate (as nitrogen) at all lake stations were consistently below the LRL (less than 0.016 mg/L).

Median concentrations of dissolved orthophosphate (as phosphorus) at all inflow stations (some of which were sampled only during storm events) and lake stations were less than 0.006 mg/L. Maximum storm inflow concentration (0.11 mg/L) occurred at Maumelle River at Williams Junction.

Median concentrations of total phosphorus at inflow stations (some of which were sampled only during storm events) ranged from 0.018 to 0.058 mg/L. Maximum storm inflow concentration (0.32 mg/L) occurred at Reece Creek at Little Italy. Median concentrations of total phosphorus at lake stations ranged from 0.011 to 0.018 mg/L. Median concentrations generally were higher in the upstream parts of the lake and decreased in the downstream parts of the lake.

Median concentrations of dissolved organic carbon (as carbon) of inflow stations (some of which were sampled only during storm events) ranged from 3.4 to 6.1 mg/L. Maximum storm inflow concentration (15 mg/L) occurred at Yount Creek near Martindale. Median concentrations of dissolved organic carbon (as carbon) in Lake Maumelle ranged from 3.1 to 3.6 mg/L. Concentrations generally were higher in the upstream parts of the lake but did not decrease consistently in the downstream direction.

Median concentrations of total organic carbon at inflow stations (some of which were sampled only during storm events) ranged from 4.3 to 8.9 mg/L (as carbon). Maximum storm inflow concentration (24 mg/L) occurred at Maumelle River at Williams Junction. Median concentrations of total organic carbon at lake stations ranged from 3.8 to 4.2 mg/L and generally decreased in the downstream parts of the lake.

Median concentrations of fecal coliform bacteria at inflow stations ranged from 77 to 2,450 colonies per 100 milliliters (col/100 mL). Maximum storm inflow concentration (12,000 col/100 mL) occurred at Reece Creek at Little Italy. Median concentrations of fecal coliform bacteria at lake stations ranged from less than 1 to 8 col/100 mL. Median concentrations generally were slightly greater in the upstream parts of the lake.

Median concentrations of suspended sediment of inflow stations (some of which were sampled only during storm events) ranged from 9 to 55 mg/L. Maximum storm inflow concentration (773 mg/L) occurred at Reece Creek at Little Italy.

Median concentrations of chlorophyll a at lake stations ranged from 3.2 to 4.0 micrograms per liter (μg/L). Concentrations varied throughout the lake with no discernible pattern.

Transparency of the lake as indicated by median Secchi disk depth of visibility below the surface ranged from 3.5 to 6.9 ft. Transparency depth was shallowest in the upstream parts of the lake and deepest in the downstream parts of the lake.

Description of Lake Maumelle Aging and Trophic Status

Reservoirs are formed by damming or impounding free-flowing streams and rivers, permanently flooding the river valley. Reservoirs reach trophic equilibrium relatively rapidly after their basins fill with water, and trophic status is determined initially by the nature of the drainage basin (Kimmel and Groeger, 1986). In reservoirs, a highly productive period, termed the "trophic upsurge" (Baranov 1961), occurs prior to the establishment of a trophic equilibrium. The "trophic upsurge" is followed by a "trophic depression," which is, in fact, the initial approach of the reservoir ecosystem toward its natural equilibrium level (Kimmel and Groeger, 1986) (fig. 3).

The trophic upsurge in new reservoirs (Kimmel and Groeger,1986) is the result of a combination of several factors: (1) a large influx of organic detritus and inorganic nutrients from the inundated reservoir basin, (2) an abundance of high quality habitat and food for benthic organisms, and (3) a rapidly expanding lacustrine environment (Baxter 1977; Ploskey 1981; Benson 1982). Similarly, the subsequent decline in biological productivity, the "trophic depression," also has multiple causes: (1) decreased internal nutrient loading, (2) a decline in biologically labile organic detritus, (3) the cessation of habitat expansion, and (4) a reduction of favorable habitat. The magnitude and duration of the trophic disequilibrium phase (fig. 3) are quite variable among reservoirs because of differing basin inundation rates, internal and external nutrient loading rates, flushing rates, the quality and quantity of new habitat, the fish assemblages present, and reservoir operations (Ploskey, 1981; Kimmel and Groeger, 1986).

The brief initial period of trophic disequilibrium characteristic of new impoundments yields to a less productive but potentially more stable period of trophic equilibrium as internal nutrient loading decreases (see 1 in fig. 3). Human-induced alterations of the watershed may cause increases in external nutrient loading and biological productivity (see 2 in fig. 3). Following the dynamic environmental and biological fluctuations characteristic of the early years of a reservoir's existence, the magnitude and variability of biological production within the maturing reservoir become dependent on inputs of nutrients and organic matter from the watershed, as in natural lakes (Kimmel and Groeger, 1986). Because of large drainage basins relative to reservoir surface areas and volumes, fluvial inputs are the most important sources of nutrients for most reservoir ecosystems (Gloss and others, 1980; Kimmel and others, 1990; Kimmel and Groeger, 1986). In addition to possible point sources of nutrients (industries and municipalities), patterns of land use within watersheds are primary determinants of the nutrient loading to aquatic systems and, thereby, of lake and reservoir productivity (Likens, 1972; Hutchinson, 1973). Undisturbed terrestrial ecosystems are usually characterized by runoff with low concentrations in dissolved substances; however, pastures, croplands, and urban areas contribute much greater nutrient loads to aquatic systems (Likens, 1975). Therefore, land-use patterns will have long-term effects on reservoir productivity and water quality (Kimmel and Groeger, 1986).

If reservoirs are permitted to age without being otherwise disturbed, one would expect (based on present understanding of the relations between basin morphology, nutrient loading rates, and lacustrine productivity) that reservoir productivity would remain relatively constant over time (for reservoirs that fill slowly from siltation) or gradually increase as mean depth decreases (for reservoir basins undergoing rapid siltation). Because construction of reservoirs (manmade impoundments) often promote additional land-use changes and technological development within reservoir watersheds and their relatively large watersheds focus both point and diffuse sources of nutrients into reservoir basins, water quality and productivity

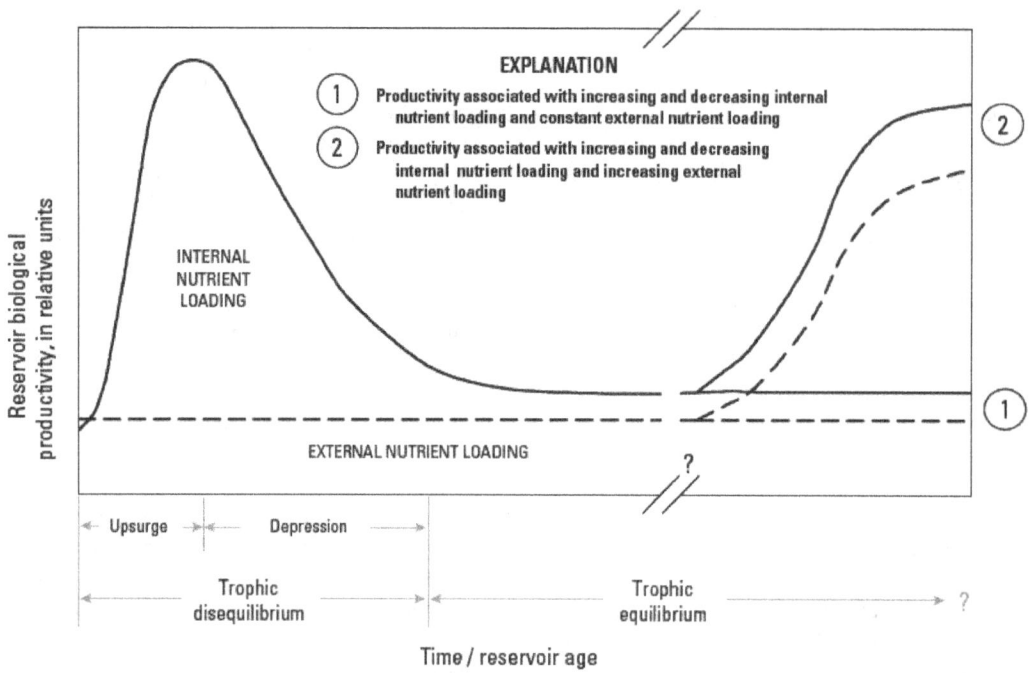

Figure 3.   Conceptual model showing changes in factors influencing reservoir water quality and biological productivity as a reservoir matures and ages.

changes attributable to "natural" reservoir aging will be small compared to the effects of human-induced changes on watershed-reservoir interactions (Kimmel and Groeger, 1986).

Lake Maumelle was completed in 1958 and has been in existence for 54 years (as of 2012). Given the reservoir aging process described above, Lake Maumelle has since passed the trophic disequilibrium phases resulting from impoundment ("trophic upsurge" and the following "trophic depression") and currently coexists in equilibrium with its watershed (external loading) where productivity has remained relatively constant over time.

Secchi disk depth and concentrations of total phosphorus and chlorophyll *a* varied little over the 2004 through 2010 period used to examine the effects of simulated land-use changes on water quality of Lake Maumelle, indicating that Lake Maumelle is within the trophic equilibrium period (fig. 3). For the model period of 2004–10, as described in this report, measures of Secchi disk depth and water-quality samples collected and analyzed for total phosphorus and chlorophyll *a* concentrations were used to calibrate the CE–QUAL–W2 model at three sites on Lake Maumelle and also were used to examine the effects of simulated land-use changes on water quality of Lake Maumelle. Median

(1989–2010) Secchi disk depths at the three sites from upper to lower end (Lake Maumelle east of Highway 10, Lake Maumelle near Little Italy, and Lake Maumelle at Natural Steps) were 4.3, 6.6, and 6.9 ft, respectively (table 2). Median total phosphorus concentrations from the upper to lower end, including all sample depths, were 0.016, 0.012, and 0.011 mg/L, respectively (table 2). Median chlorophyll *a* concentrations from the upper to lower end, 3 ft below the surface, were 3.2, 3.5, and 3.3 µg/L, respectively (table 2).

The ratio of chlorophyll *a* and total phosphorus can be used to predict whether reservoir water quality can be improved by reducing the external phosphorus load, or alternatively, the consequences of its increase (Straskraba and others, 1993). The relation between the two characteristics usually is considered to be exponential, increasing rapidly at low concentrations (Dillon and Rigler, 1974). When total phosphorus reaches a certain (saturated) concentration, the increase in chlorophyll *a* will stop increasing and the relation will be sigmoid (fig. 4) in the whole range of values (Straskraba and others, 1993; Straskraba, 1976, 1978). This relation has been confirmed in other studies (Straskraba, 1985; Prairie and others, 1989). Phytoplankton concentrations, expressed as concentrations of chlorophyll *a*,

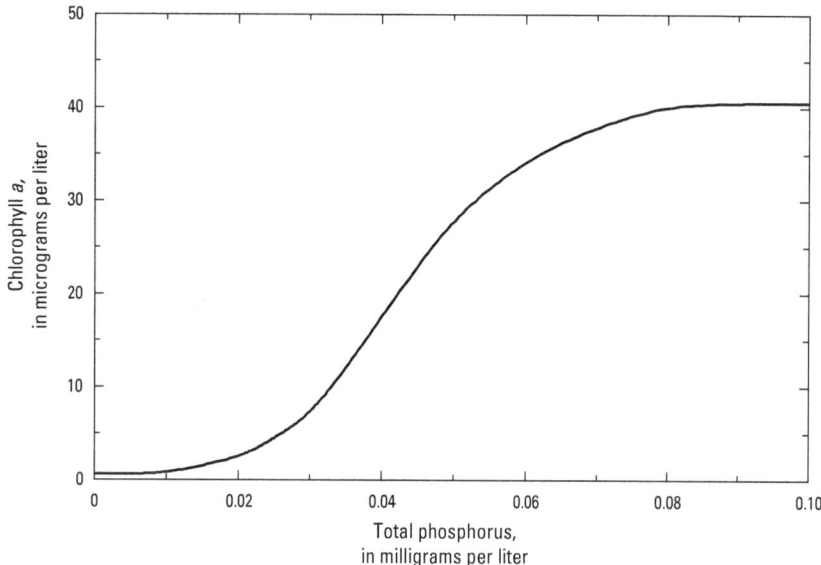

Figure 4. Relation between chlorophyll *a* concentration and total phosphorus concentration in lakes and reservoirs.

are not very sensitive to total phosphorus concentrations up to approximately 0.020 mg/L; its response is very strong between approximately 0.020 and 0.060 mg/L (Straskraba and others, 1993). However, not all reservoirs are phosphorus limited. At the weight ratio of total nitrogen to total phosphorus (TN:TP) less than 10, phytoplankton appear to be limited by nitrogen. During nitrogen limitation, the effect of a change in nitrogen concentration on chlorophyll *a* can be roughly estimated in figure 4 by plotting the value of one-tenth the total nitrogen concentration on the x-axis (Straskraba and others, 1993).

Median total phosphorus concentrations in Lake Maumelle for the 1989 to 2010 period described above, ranged from 0.011 to 0.016 mg/L at the three monitoring (calibration) sites and varied little, whereas the median chlorophyll *a* concentrations ranged from 3.2 to 3.5 µg/L. Given these positions on the sigmoid total phosphorus and chlorophyll *a* plot (fig. 4), the conditions in Lake Maumelle exist below the "very sensitive" range described by Straskraba and others (1993) bounded by total phosphorus concentrations between 0.020 and 0.060 mg/L. The 0.020 mg/L threshold is not exact; however, the ambient total phosphorus concentrations in Lake Maumelle are not far below this threshold.

The trophic response variables and other water-quality constituents used to calibrate the CE–QUAL–W2 model for the period 2004–10 reflect the stable trophic equilibrium phase in reservoir aging described by Kimmel and Groeger (1986) (fig. 3). As such, and because the CE–QUAL–W2 model

is bound by the range in data that are used to calibrate the hydrodynamic and water-quality processes used to examine the effects of simulated land-use changes on water quality in Lake Maumelle, model results will be limited and thus defined by these trophic boundaries.

Hydrologic Simulation Program–FORTRAN Watershed Modeling Development

HSPF is a continuous watershed model developed to simulate the hydrologic and associated water-quality processes on a specified time step for pervious and impervious land surfaces and in streams (Bicknell and others, 2001). HSPF is a semilumped-parameter model that simulates spatial variability by discretizing the watershed into homogeneous land units based on relative similarity of land use, soils, topography, and other hydrologic characteristics. A water budget is calculated by HSPF where inflows equal outflows plus or minus change in storage for each time step. Within HSPF, some parameters are measured or default values; however, most are adjusted during the calibration process. A complete description of the mathematical equations and model variables can be found in Bicknell and others (2001).

Numerous datasets are required in the development of a HSPF model. Datasets compiled for this study include the National Elevation Dataset (NED) (U.S. Geological Survey, 2009b) for use in determining hydrologically similar land areas; the National Hydrography Dataset (NHD) (U.S. Geological Survey, 2009c) that includes all the stream reaches within the Lake Maumelle watershed; the 2006 Arkansas land-use/land-cover map (Arkansas Natural Resources Commission and University of Arkansas: Center for Advanced Spatial Technologies, 2009); aerial photography taken February 2009 (Martin Maner, Central Arkansas Water, written commun., 2011); the Soil Survey Geographic (SSURGO) database for each county within the watershed (U.S. Department of Agriculture, 2009); and Next-Generation Radar (NEXRAD) hourly precipitation data (National Climatic Data Center, 2008), as well as other meteorological data including air temperature, solar radiation, dewpoint temperature, wind velocity, and cloud cover. Air temperature, dewpoint temperature, wind velocity, and cloud cover were obtained from the National Climatic Data Center (NCDC) stations surrounding the watershed and include Hot Springs Memorial Field Airport, Russellville Municipal Airport, and Little Rock Air Force Base (National Oceanic and Atmospheric Administration, 2011) (fig. 5).

Five streamflow-gaging stations (fig. 6) were used in the calibration process and included Maumelle River at Williams Junction (07263295), hereafter referred to as Williams Junction; Maumelle River near Wye (07263296), hereafter referred to as Wye; Bringle Creek at Martindale (072632962), hereafter referred to as Bringle Creek; Yount Creek near Martindale (072632971), hereafter referred to as Yount Creek; and Reece Creek at Little Italy (072632982), hereafter referred to as Reece Creek. Continuous streamflow data with varying lengths of record are available for each station (table 1). Williams Junction is the only station with continuous streamflow data for the entire HSPF model simulation period. Streamflow records for each station were retrieved from the USGS National Water Information System (U.S. Geological Survey, 2010). During dry periods, Reece, Yount, and Bringle Creeks have no streamflow; these periods generally occurred during the months of June, July, August, and September.

Along with streamflow, selected water-quality field measurements and various constituent concentrations were used in the calibration process. At each station, water temperature and dissolved oxygen were measured, and samples were collected for analysis of suspended sediment, fecal coliform bacteria, dissolved nitrite plus nitrate nitrogen, dissolved ammonia nitrogen, dissolved orthophosphate, total phosphorus, and total organic carbon. Samples generally were collected quarterly or monthly and during selected storm events at the two Maumelle River stations of Williams Junction and Wye. Reece, Yount, and Bringle Creeks were sampled only during selected stormflow events (streamflow substantially influenced by runoff). The events at Reece, Yount, and Bringle Creeks were sampled with automated water-quality samplers; each station could have had more than

one sample collected on any one particular day. Additionally, equal-width-increment samples (U.S. Geological Survey, 2006) were collected during approximately three storm events per year. For most constituents, samples were collected on 54 days at Wye, 65 days at Williams Junction, 54 days at Reece Creek, 45 days at Bringle Creek, and 60 days at Yount Creek during the HSPF model simulation period, and each sample was considered for calibration.

Subwatershed Delineation and Land Use

HSPF requires a set of hydrologically similar land areas that compose the watershed. These hydrologically similar areas are grouped into subwatersheds containing pervious land area (PERLND) and impervious land area (IMPLND), each of which simulates the water quality and quantity processes that occur on the land and were determined from spatial land-use datasets. Subwatersheds (fig. 6) were delineated based on NED and automatic delineation tools within a Geographic Information System (GIS) and were adjusted manually based on the NHD stream reaches and CAW land ownership maps. There are 55 subwatersheds in the HSPF model, characterizing Lake Maumelle's watershed, covering approximately 136 mi^2 (86,907 acres). Several subwatersheds were created using a 0.25-mi buffer around the lake; most of the area within these subwatersheds was assumed to be undeveloped or anthropogenically unaltered because of CAW ownership. Discretizing the CAW owned land into individual subwatersheds allowed for specific control of the parameters associated with the subwatersheds and also control of the input from these subwatersheds into the CE–QUAL–W2 model. For example, the Reece Creek watershed consists of subwatersheds 23, 24, and 25, and therefore, subwatershed 25, adjacent to the lake, represents the simulated outflow and water quality for Reece Creek from the HSPF model (fig. 6). Each subwatershed and its associated PERLND and IMPLND drains into a stream segment or reach reservoir (RCHRES). There is one RCHRES per subwatershed and each RCHRES subsequently drains downstream. Land use was determined from the Arkansas 2006 land-use/land-cover dataset (Arkansas Natural Resources Commission and University of Arkansas: Center for Advanced Spatial Technologies, 2009) and the aerial photography flown (contracted) by CAW in early 2009. The land use was categorized into eight PERLND types and seven IMPLND types (table 3). IMPLNDs were assigned as a percentage of the PERLND based on the land-use/land-cover datasets.

Pervious and Impervious Land Segments

The water-quality and quantity processes that occur on the land surface before entering a stream segment are simulated within each PERLND or IMPLND module. The processes that occur within the IMPLNDs include only surface runoff, no infiltration occurs; whereas, the processes that occur within the PERLNDs include water

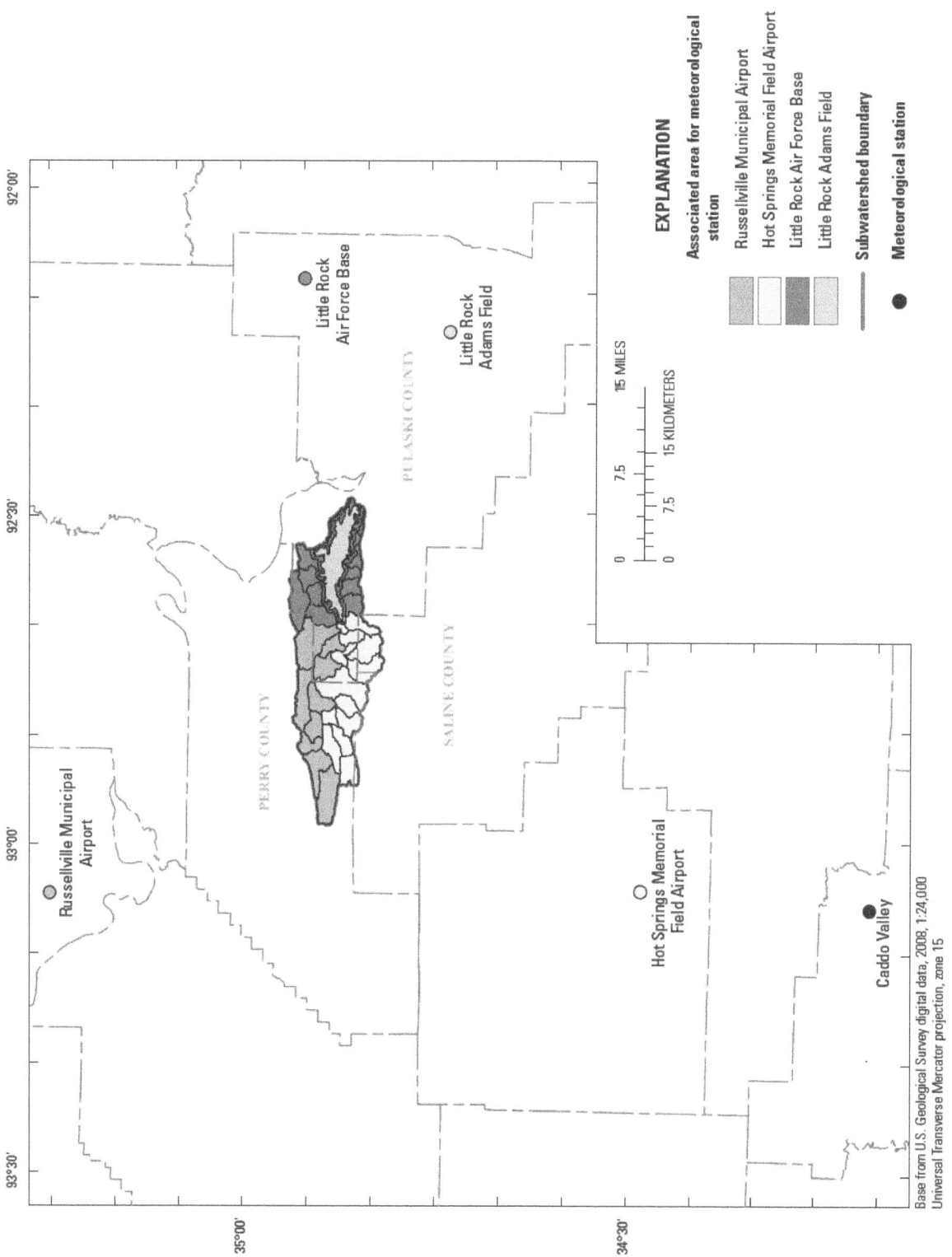

Base from U.S. Geological Survey digital data, 2008, 1:24,000
Universal Transverse Mercator projection, zone 15

Figure 5. Locations of meteorological stations and associated areas within the Lake Maumelle watershed used for the Hydrologic Simulation Program–FORTRAN model and the CE–QUAL–W2 model.

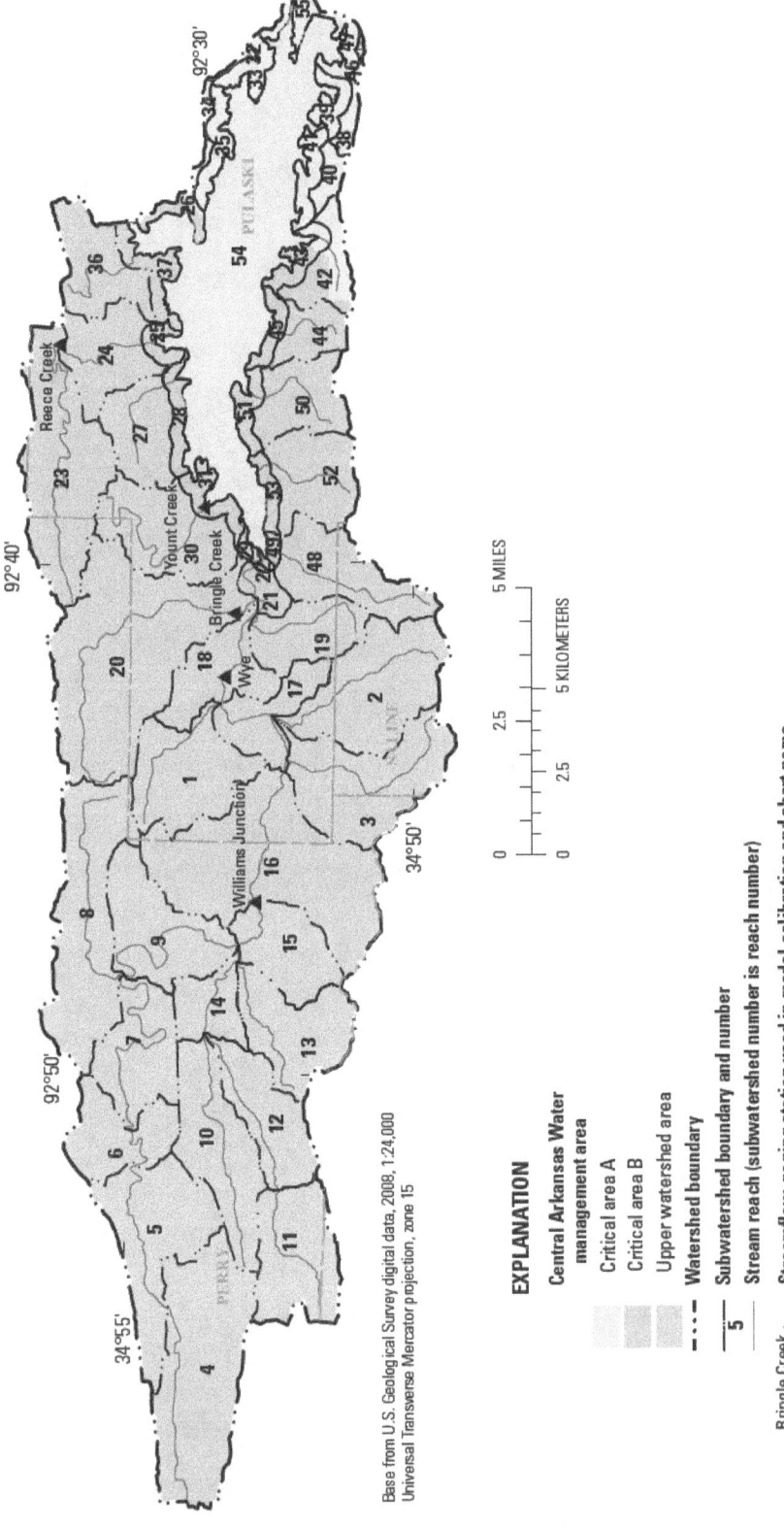

Base from U.S. Geological Survey digital data, 2008, 1:24,000
Universal Transverse Mercator projection, zone 15

EXPLANATION

**Central Arkansas Water
management area**

    Critical area A

    Critical area B

    Upper watershed area

- - - · ·   Watershed boundary

————   Subwatershed boundary and number
   5

————   Stream reach (subwatershed number is reach number)
   5

Bringle Creek ▲   Streamflow-gaging station used in model calibration and short name

Figure 6.   Locations of streamflow-gaging stations, subwatersheds and reaches, and Central Arkansas Water management areas.

Table 3. Land-use types and characteristics designated for pervious and impervious areas within subwatersheds, Lake Maumelle, Arkansas.

[Percentage does not total to 100 percent because of rounding; PERLND, pervious land area; IMPLND, impervious land area; --, not designated]

Land use	PERLND type number	IMPLND type number	Land-use percentage	Total area (square miles)
Agriculture	3	1	0.05	0.07
Bare soil	8	5	0.06	0.08
Clearcut	5	4	5.60	7.61
Coniferous forest	4	3	56.79	77.12
Deciduous forest	7	3	23.11	31.39
Grasslands	9	8	3.06	4.16
Paved roads	2	2	0.79	1.07
Urban	1	7	0.39	0.53
Water (Lake Maumelle)	--	--	10.14	13.77

movement by three components: surface runoff, interflow, and groundwater. Runoff is simulated for each PERLND or IMPLND independently, and a water balance is calculated for each time step throughout the simulation. Surface runoff associated with the IMPLNDs and PERLNDs is routed into a stream reach within the associated subwatershed. Interflow and groundwater within the PERLNDs account for water that is not evaporated or moved off the land surface from direct runoff. Interflow accounts for water that is directly infiltrated to the upper soil zone, is moved to overland flow from surface storage, or is removed from interflow storage to shallow subsurface flow. Groundwater accounts for the water that infiltrates to the lower soil zone and groundwater storage. The water that is retained as groundwater storage reappears as base flow or is lost from the system through deep percolation.

Reach Characterization

Within each subwatershed, water and associated water-quality constituents drain from the PERLNDs and IMPLNDs into a RCHRES. Each stream reach (fig. 6) within the HSPF model is characterized by a piecewise linear function table (FTABLE) and is developed based on channel geomorphology, width, depth, length, slope, and roughness for streamflow routing. Channel characteristics, such as width, depth, length, and slope (roughness was estimated), for each reach were determined from cross-sectional information

provided in the USGS National Water Information System or from GIS. These data were then input into a tool provided by Aqua Terra Consultants (Brian Bicknell, written commun., 2011) to create the FTABLEs. FTABLEs are independent of the shape of the water body, but they serve to relate stage to surface area, channel volume, and discharge and are based on the one-dimensional kinematic wave theory for each stream reach (Moore and Mohamoud, 2007; Bicknell and others, 2001). The assumptions made to calculate water movement through a reach are: flow within a stream reach is assumed to be well-mixed and unidirectional; inflows enter a stream reach at its upstream limit; and precipitation, evaporation, and fluxes from the PERLNDs and IMPLNDs influence processes that occur within the stream reach.

Meteorological Data

The sparseness of available rain-gage data in small watersheds is a severe hindrance to accurate hydrologic modeling. As part of its NEXRAD program, the National Weather Service (NWS) River Forecasting Centers (RFCs) produce gridded precipitation estimates; these estimates are known as Multisensor Precipitation Estimator (MPE) data (National Oceanic and Atmospheric Administration, 2010). MPE data are based from Doppler radar precipitation data and replace the earlier Stage III NEXRAD product (National Climatic Data Center, 2008). The MPE algorithms provide better gage-correction biasing, mosaicking of radar data, and can incorporate satellite-derived precipitation estimates into the final MPE data product (National Oceanic and Atmospheric Administration, 2002).

The MPE data offer precipitation estimates spatially averaged over grid cells of about 6 mi^2 and temporally averaged over 1 hour (fig. 7). The Weather Surveillance Radar – 1998 Doppler (WSR–88D) weather radar that provides raw radar data for the Lake Maumelle watershed is located approximately 25 mi east of the watershed. A single WSR–88D beam has an effective range of approximately 143 mi (U.S. Army Corps of Engineers, 1994). The NEXRAD products provide hourly estimates in the Hydrologic Rainfall Analysis Project (HRAP) grid system, about a 2.5-mi grid in a Polar Stereographic map projection (Shedd and Fulton, 1993). The HRAP grid (Greene and Hudlow, 1982) is used to identify the location of each NEXRAD precipitation value. This spatially distributed precipitation data can be incorporated into watershed models as an improvement to using the sparse rain-gage networks to obtain rainfall data (Ockerman and Roussel, 2009; Soong and others, 2005). MPE data were used as the precipitation input for HSPF and CE–QUAL–W2 models.

MPE data for the model simulations were obtained from the Lower Mississippi RFC (National Oceanic and Atmospheric Administration, 2008). The HRAP grid cells with the hourly MPE precipitation time series were intersected with each of the subwatersheds of the HSPF model using standardized functions within a GIS. The amount of

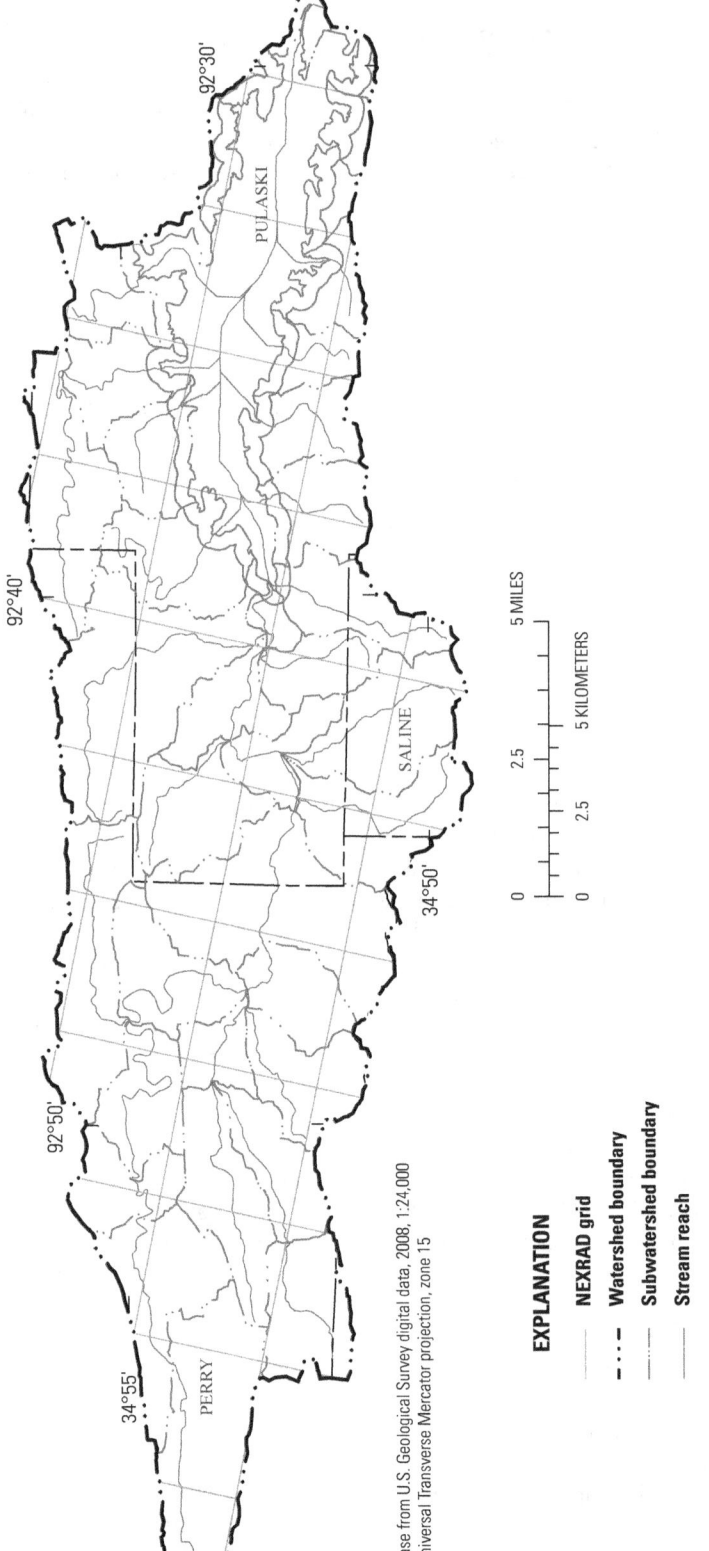

Base from U.S. Geological Survey digital data, 2008, 1:24,000
Universal Transverse Mercator projection, zone 15

EXPLANATION

——— NEXRAD grid

—··—·· Watershed boundary

——···—— Subwatershed boundary

——— Stream reach

Figure 7.   Location of the Next Generation Radar (NEXRAD) grid, Lake Maumelle watershed, Arkansas.

precipitation each subwatershed received was determined by weighting the percentage of subwatershed area covered by each HRAP grid cell. For example, if the entire area of one subwatershed is within one HRAP grid cell, the subwatershed would receive a proportionate amount of precipitation equivalent to the amount of HRAP grid cell area covered by the subwatershed; if a subwatershed is split by more than one HRAP grid cell, each grid cell precipitation value is multiplied by the percentage of subwatershed area that falls within each HRAP cell, and then all products are summed together.

During the model simulation time period of January 2004 through December 2010, the fifth driest year (2005) on record (1895–2010), with only 36.2 inches, and the wettest year (2009) on record, with 72.7 inches, occurred. The average annual precipitation for Arkansas from 1895–2010 was 49.5 inches (National Oceanic and Atmospheric Administration, 2012).

Atmospheric deposition of ammonia and nitrate was applied as a time series to the HSPF model. Concentrations of ammonia and nitrate were input into the HSPF model as hourly values, disaggregated from mean monthly atmospheric deposition concentrations at Caddo Valley meteorological station (National Atmospheric Deposition Program, 2012) (fig. 5) using a set of preprocessors developed by the U.S. Environmental Protection Agency (2003), and distributed during precipitation events to all PERLNDs and RCHRESs.

Solar radiation measured in langleys was input as hourly values and was computed from measured percent cloud cover and latitude of the watershed. The values were derived using a set of preprocessors developed by the U.S. Environmental Protection Agency (2003).

Air temperature, wind speed, dew point, and cloud cover were obtained from the NCDC (National Oceanic and Atmospheric Administration, 2011). Three meteorological stations surrounding the watershed include Hot Springs Memorial Field Airport, Russellville Municipal Airport, and Little Rock Air Force Base (fig. 5). Values for these meteorological characteristics were distributed to subwatersheds dependent on proximity to the station (fig. 5). Cloud cover was input as daily values; air temperature was input as hourly values derived using daily minimum and maximum temperatures; dew point temperatures were input as hourly values derived using monthly values; wind speed was input as daily values derived using monthly values; and potential evapotranspiration estimates were determined from minimum and maximum daily temperature, as well as watershed latitude using the Hamon method (Hummel and others, 2001). Meteorological data were obtained as daily or monthly values and, when necessary, were disaggregated using a preprocessor (U.S. Environmental Protection Agency, 2003) into hourly or daily data. If the data received from NCDC were missing or were on varying intervals, a mean was obtained for the day or month and then was disaggregated into hourly or daily data depending on the needs of the model.

CE–QUAL–W2 Reservoir Modeling Development

A two-dimensional, laterally averaged, hydrodynamic and water-quality model (using CE–QUAL–W2 Version 3.6 [Cole and Wells, 2008]) was developed for Lake Maumelle and calibrated using data collected during January 2004 through December 2010. The model simulates 14 active constituents (temperature, dissolved solids, inorganic suspended solids, two nitrogen and one phosphorus constituent, iron, phytoplankton [total algae] and epiphyton, labile and refractory dissolved and particulate organic matter, and dissolved oxygen) and 6 derived constituents (dissolved organic carbon, total organic carbon, ammonia plus organic nitrogen, total nitrogen, total phosphorus, chlorophyll *a*). The model uses a variable time step, with a minimum time step of 1 second. Complete details of model theory and structure and an extensive bibliography for theoretical development and application are given by Cole and Wells (2008).

Development of the CE–QUAL–W2 model of Lake Maumelle included the computational grid, specification of boundary and initial conditions, and preliminary selection of model parameter values. The CE–QUAL–W2 model development and associated assumptions in the selection of boundary and initial conditions are described, and model parameters are listed in this section.

Computational Grid

The CE–QUAL–W2 model domain extends from west of the East of Highway 10 station (fig. 1) to the Lake Maumelle Dam and Spillway (fig. 1), a distance of approximately 12 mi along the longitudinal axis of the reservoir. The model grid segments representing the coverage of the lake were broken into 28 computational segments (fig. 8). Segments 2 and 22 represent the farthest upstream and downstream parts of the lake, respectively; two side branches on the lake were represented by segments 25 through 28 and 31 through 33, and segments 1, 23, 24, 29, 30, and 34 were inactive boundary segments. Unaccounted flow in the tributaries was distributed equally into each segment through the "distributed tributary" inflow file. Segments ranged in length from 425 to 1,372 yards (yds) and in surface width at spillway altitude from 683 to 3,460 yds, within the main water-body branch. Each segment was divided vertically into 3.28-ft (1 m) layers; therefore, the lake's depth and width were allowed to vary spatially. The orientation of the longitudinal axis of each segment relative to north was determined for each segment. This information was used in the computation of surface wind stress in each segment. Grid geometry and reservoir bathymetry were

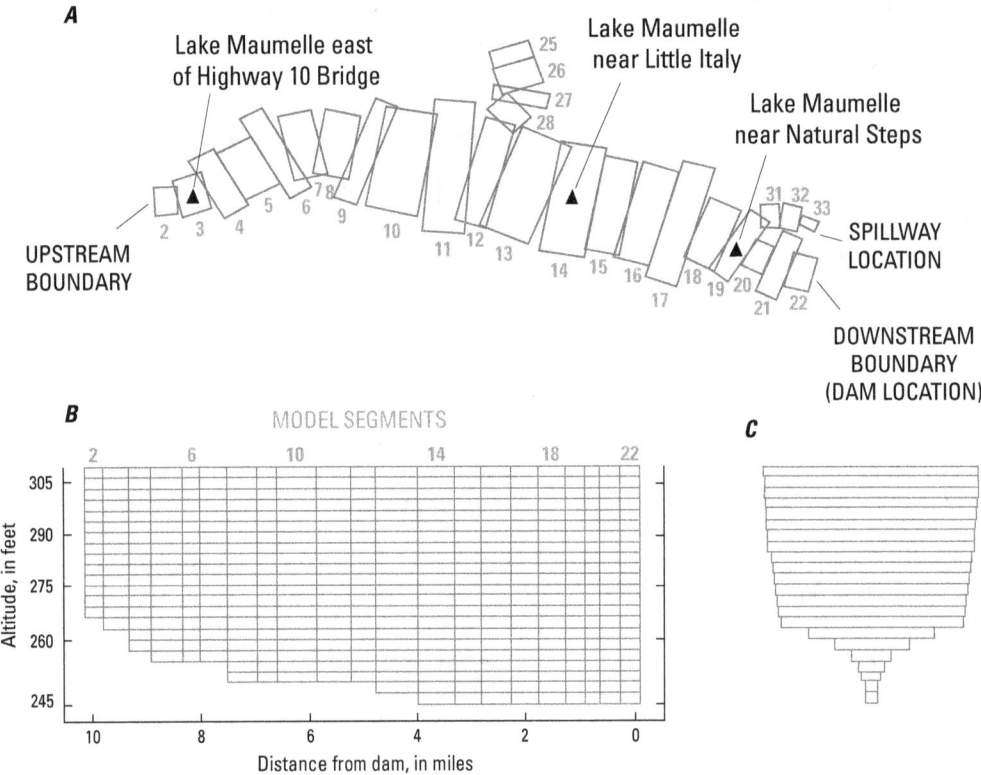

Figure 8. The CE–QUAL–W2 model framework, Lake Maumelle watershed, Arkansas.

developed using preimpoundment altitude contours of the reservoir bottom from the original engineering land surveys provided by CAW. The preimpoundment altitude contours for Lake Maumelle were digitized by the USGS during the course of this project to improve the resolution and accuracy of the lake model.

Boundary and Initial Conditions

Hydraulic, thermal, and chemical boundary conditions are required for a CE–QUAL–W2 model (Cole and Wells, 2008). The boundaries of the Lake Maumelle model included the reservoir bottom, the shoreline, tributary streams, the upstream boundary, the downstream boundary, and the water-surface altitude of the reservoir. Initial water-surface altitude of the reservoir, water temperature, and selected water-quality constituent concentrations (based on January 2003 conditions) also are required.

Hydraulic and Thermal Boundary Conditions

The reservoir bottom was assumed to be an immobile and impermeable boundary within the CE–QUAL–W2 model. That is, the bottom sediments were stationary and not resuspended by flow or groundwater discharge to the reservoir, and recharge from the reservoir to groundwater was negligible. The reservoir bottom extracts energy from water movement by causing resistance to water flow; this phenomenon varied with the magnitude of flow. A single, empirical coefficient (Chezy resistance coefficient) was applied to the reservoir bottom in all computational segments to simulate effects of bottom friction.

Heat exchange between the reservoir bottom and overlying water was computed from (1) the reservoir bottom (sediment) temperature and (2) a coefficient of bottom heat exchange (table 4). The reservoir bottom temperature and coefficient of bottom heat exchange were assumed to be constant in space and time. Heat exchange at the reservoir bottom is typically quite small (about two orders of magnitude less than surface heat exchange) (Cole and Wells, 2008).

The reservoir shoreline is defined as a boundary across which there is no flow. The exact position of the shoreline changes during a model simulation because of changing water surface.

Discharge over the spillway (290 ft above NGVD) was monitored by a stage recorder. The simulated spillway was positioned in segment 33 (fig. 8). Daily drinking-water-supply withdrawal was monitored by CAW. The simulated withdrawal structure was positioned at an altitude of 269 ft in segment 22 (fig. 8).

Hydraulic boundary conditions and internally calculated values at the water surface included precipitation, wind stress, and surface heat exchange. All meteorological data, with the exception of precipitation, required for these computations were measured at Little Rock Adams Field (fig. 5), about 18 mi southeast of the Lake Maumelle Dam (fig. 1), and generally were recorded at hourly intervals.

Other boundary conditions included precipitation on the pool (quantity and quality) and meteorology. Precipitation on the reservoir water surface was based on NEXRAD rainfall data (see NEXRAD discussion within the "Meteorological Data" section for further information). Evaporation was computed in the CE–QUAL–W2 model from a time series of water-surface temperature, dewpoint temperature, and area of the exposed lake surface. Wind stress was computed from a time series of wind speed and direction, the orientation of the model grid segment, and a wind-sheltering coefficient. The wind-sheltering coefficient reduces the effects of wind on the reservoir because of topographic sheltering of the water surface. A term-by-term accounting was used for surface heat exchange using latitude and longitude provided in the control file, values for air and dewpoint temperature, wind speed and direction, and cloud cover provided in the meteorological file.

Chemical Boundary Conditions

In addition to temperature, concentrations of the following constituents were simulated for Lake Maumelle: labile dissolved organic matter, refractory dissolved organic matter, algae, particulate organic matter, nitrite plus nitrate nitrogen, ammonia nitrogen, dissolved orthophosphate, total phosphorus, and dissolved oxygen. Secchi disk depths were derived by incorporating the following CE–QUAL–W2 simulated constituent concentrations: inorganic suspended solids, particulate organic matter, and total algae. A time series of concentrations of selected constituents at all inflow boundaries are required for the CE–QUAL–W2 model operation; however, boundary conditions are not required for all constituents.

Inflow chemical boundary conditions for the Maumelle River at the upstream end of the lake, Reece Creek, Yount Creek, and the distributed tributaries surrounding the lake were based on calibrated output values from the HSPF model. Hourly labile and refractory dissolved organic matter and labile and refractory particulate organic matter concentrations were derived by portioning out hourly HSPF total organic carbon load into the four components. Hourly nitrite plus nitrate nitrogen, ammonia nitrogen, and dissolved orthophosphate concentrations were computed from hourly HSPF nitrite plus nitrate nitrogen, ammonia nitrogen, and dissolved orthophosphate loads. Hourly dissolved oxygen concentrations from HSPF were input into CE–QUAL–W2. Organic matter, phosphorus, and iron inputs from the reservoir bottom boundary were derived within the model and were based on the value of selected parameters (table 4) and the concentration of the constituent and the oxygen concentration in the overlying waters.

Atmospheric wet deposition of ammonia and nitrate was applied to the surface of Lake Maumelle as a time series to the CE–QUAL–W2 model. Concentrations of ammonia and nitrate were input into the model as daily values, disaggregated from weekly atmospheric deposition concentrations at Caddo Valley (National Atmospheric Deposition Program, 2012) (fig. 5). Ammonia and nitrate inputs were distributed across the entire surface of Lake Maumelle based on the interpolated concentrations and volume of the precipitation event (from NEXRAD data). Labile dissolved organic matter (LDOM) was input similarly at a constant concentration of 4.44 mg/L (Gaffney and Marley, 2010).

Initial Conditions

Initial water-level altitude, water temperature, and constituent concentrations for each CE–QUAL–W2 model segment are required at the start of a CE–QUAL–W2 model simulation. Initial water-level altitudes were set to the measured value on January 1, 2003. Lake Maumelle was assumed to be isothermal (48.0°F) throughout the entire reservoir. Initial constituent concentrations also were assumed to be uniform based on concentrations measured at the Natural Steps station (fig. 1).

Model Parameters

Parameters are used to describe physical and chemical processes that are not explicitly modeled and to provide chemical kinetic rate information. Many parameters cannot be measured directly and are often adjusted during the model calibration process until model simulations agree with measureable parameters.

Table 4.   Parameters and values used in the CE-QUAL-W2 model of Lake Maumelle, January 2004 to December 2010.

[Numbers in bold are CE-QUAL-W2 default values (Cole and Wells, 2008); ---, not used]

Parameter description	Name	Values	Units
Hydraulic and thermal input parameters			
Coefficient of bottom heat exchange	CBHE	0.3	watts/square meter/second
Sediment temperature	TSED	16.5	degrees Celsius
Wind-sheltering coefficient	WSC	0.8	dimensionless
Horizontal eddy viscosity	AX	**1.0**	square meters/second
Horizontal eddy diffusivity	DX	**1.0**	square meters/second
Rate coefficients for water-chemistry and biological simulations			
Light extinction coefficient for pure water (λ_{H2O})	EXH2O	**0.45**	1/meter
Light extinction coefficient for organic solids (particulate organic matter; ε_{POM})	EXOM	**0.10**	1/meter
Light extinction coefficient for inorganic solids (inorganic suspended solids; ε_{ISS})	EXSS	**0.10**	1/meter
Light extinction coefficient because of algae	EXA	**0.20**	1/meter
Fraction of incident solar radiation absorbed at water surface	BETA	0.55	dimensionless
Suspended solids settling rate	SSS	0.2	meters/day
Algal growth rate	AG	**2.0**	1/day
Algal mortality rate	AM	**0.1**	1/day
Algal excretion rate	AE	**0.04**	1/day
Algal dark respiration rate	AR	**0.04**	1/day
Algal settling rate	AS	**0.1**	meters/day
Saturation light intensity	ASAT	**75**	watts/square meter
Fraction of algal biomass lost by mortality to particulate organic matter	ALPOM	**0.8**	dimensionless
Lower temperature for algal growth	AT1	**5.00**	degrees Celsius
Fraction of algal growth at lower temperature	AK1	**0.10**	dimensionless
Lower temperature for maximum algal growth	AT2	15.00	degrees Celsius
Fraction of maximum algal growth at lower temperature	AK2	**0.99**	dimensionless
Upper temperature for maximum algal growth	AT3	**30.00**	degrees Celsius
Fraction of maximum algal growth at upper temperature	AK3	**0.99**	dimensionless
Upper temperature for algal growth	AT4	**40.00**	degrees Celsius
Fraction of algal growth at upper temperature	AK4	**0.10**	dimensionless
Algal half-saturation constant for phosphorus	AHSP	**0.003**	grams/cubic meter
Algal half-saturation constant for nitrogen	ASHN	**0.014**	grams/cubic meter
Algal half-saturation constant for silica	AHSSI	**0.000**	grams/cubic meter
Chlorophyll-algae ratio	ACHA	0.17	dimensionless
Periphyton growth rate	EG	**2.00**	1/day
Periphyton mortality rate	EM	**0.1**	1/day
Periphyton excretion rate	EE	**0.04**	1/day
Periphyton dark respiration rate	ER	**0.04**	1/day
Periphyton saturation light intensity	ESAT	**75**	watts/square meter
Periphyton burial rate	EB	**0.1**	1/day
Fraction of peiphyton biomass lost by mortality to particulate organic matter	EPOM	**0.8**	dimensionless
Lower temperature for periphyton growth	ET1	**5.0**	degrees Celsius
Fraction of periphyton growth at lower temperature	EK1	**0.10**	dimensionless
Lower temperature for maximum periphyton growth	ET2	**25.0**	degrees Celsius
Fraction of maximum periphyton growth at lower temperature	EK2	**0.99**	dimensionless
Upper temperature for maximum periphyton growth	ET3	**35.0**	degrees Celsius

Table 4. Parameters and values used in the CE-QUAL-W2 model of Lake Maumelle, January 2004 to December 2010.—Continued

[Numbers in bold are CE-QUAL-W2 default values (Cole and Wells, 2008); ---, not used]

Parameter description	Name	Values	Units
Rate coefficients for water-chemistry and biological simulations—Continued			
Fraction of maximum periphyton growth at upper temperature	EK3	**0.99**	dimensionless
Upper temperature for periphyton growth	ET4	**40.0**	degrees Celsius
Fraction of periphyton growth at upper temperature	EK4	**0.10**	dimensionless
Peirphyton half-saturation constant for phosphorus	EHSP	**0.003**	grams/cubic meter
Periphyton half-saturation constant for nitrogen	EHSN	**0.014**	grams/cubic meter
Peirphyton half-saturation constant for silica	EHSSI	**0.000**	grams/cubic meter
Chlorophyll-periphyton ratio	ECHLA	**145**	dimensionless
Labile dissolved organic matter decay rate	LDOMDK	0.05	1/day
Refractory dissolved organic matter decay rate	RDOMDK	0.001	1/day
Labile to refractory dissolved organic matter decay rate	LRDDK	0.001	1/day
Labile particulate organic matter decay rate	LRPDK	0.08	1/day
Refractory particulate organic matter decay rate	RPOMDK	0.001	1/day
Labile to refractory particulate organic matter decay rate	LRPDK	**0.001**	1/day
Particular organic matter settling rate	POMS	0.1	meters/day
Lower temperature for organic matter decay	OMT1	5	degrees Celsius
Upper temperature for organic matter decay	OMT2	30	degrees Celsius
Fraction of organic matter decay at lower temperature	OMK1	**0.1**	dimensionless
Fraction of organic matter decay at upper temperature	OMK2	0.99	dimensionless
Sediment decay rate	SEDK	0.1	1/day
Zero-order sediment oxygen demand	SOD	0.5–2.0	grams/square meter/day
Fraction of sediment oxygen demand	FSOD	1.0	dimensionless
5-day biological oxygen demand decay rate	KBOD	0.15	1/day
Biological oxygen demand temperature rate coefficient	TBOD	---	dimensionless
Ratio of 5-day biological oxygen demand to ultimate biological oxygen demand	RBOD	---	dimensionless
Release rate of phosphorus from bottom sediment	PO4R	**0.001**	fraction of sediment oxygen demand
Phosphorus partitioning coefficient	PARTP	0.0	dimensionless
Release rate of ammonia from bottom sediment	NH4R	0.015	fraction of sediment oxygen demand
Ammonia decay rate	NH4DK	0.12	1/day
Lower temperature for ammonia decay	NH4T1	**5.0**	degrees Celsius
Fraction of nitrification at lower temperature	NH4K1	0.1	dimensionless
Upper temperature for ammonia decay	NH4T2	35	degrees Celsius
Fraction of maximum nitrification at lower temperature	NH4K1	**0.99**	dimensionless
Nitrate decay rate	NO3DK	**0.03**	1/day
Lower temperature for nitrate decay	NO3T1	**5.0**	degrees Celsius
Fraction of denitrification at lower temperature	NO3K1	**0.1**	dimensionless
Upper temperature for nitrate decay	NO3T2	20	degrees Celsius
Fraction of maximum denitrification at lower temperature	NO3K2	**0.99**	dimensionless
Iron release from bottom sediment	FER	0.5	fraction of sediment oxygen demand
Iron settling velocity, meters/day	FES	**1.0**	
Oxygen stoichiometric equivalent for ammonia decay	O2NH4	**4.57**	dimensionless
Oxygen stoichiometric equivalent for organic matter decay	O2OM	**1.4**	dimensionless
Oxygen stoichiometric equivalent for algal dark respiration	O2AR	**1.1**	dimensionless
Oxygen stoichiometric equivalent for algal growth	O2AG	**1.4**	dimensionless

Most of the hydrodynamic and thermal processes are modeled in CE–QUAL–W2, which results in very few adjustable hydraulic and thermal parameters. Many rate coefficients for water-chemistry and biological simulations are required for the application of CE–QUAL–W2 (table 4). Many of the rate coefficients were based on suggested values given as default values for CE–QUAL–W2; others were based on previous modeling applications (Haggard and Green, 2002; Galloway and Green, 2002, 2003; Green and others, 2003; Bales and others, 2001; Sullivan and Rounds, 2005; Tetra Tech, Inc., 2007).

Evaluation Methods for the Hydrologic Simulation Program–FORTRAN and CE–QUAL–W2 Models

Evaluation methods and graphical methods can be used to evaluate the acceptance criteria of model calibration. Model calibration is an iterative process of simulation, parameter evaluation, and adjustment to achieve an acceptable match of simulated values to measured values. Model validation was not performed for the HSPF or CE–QUAL–W2 models because of the limited data for calibration (2005 or later for most inflow stations). The simulation period (2004–10) included extreme dry and wet years as well as more normal hydrologic conditions. Randomly selecting one or more years from the calibration data set to use for validation would have potentially removed valuable measures needed to calibrate the model for the range of conditions. Instead, it was determined to include all of the recent calibration data in the calibration process to obtain the most accurate representative model.

Several comparisons were used to evaluate the HSPF and CE–QUAL-W2 models. Statistics used to evaluate streamflow calibration—streamflow volumes, coefficient of determination (R^2), the Mean Absolute Error (MAE), Root Mean Square Error (RMSE), and the Nash-Sutcliffe model efficiency coefficient (NSE) —were calculated for the simulation period (2004–10) for the inflow stations (table 1). HSPF statistics for water-quality characteristics included mean daily values and were calculated only for Williams Junction (because it was the station with the longest period of record and the reach with the largest contributing flow). Mean differences were calculated between daily simulated values and measured reservoir values for the three lake stations (East of Highway 10, Little Italy, and Natural Steps; fig. 1). Additionally, MAE and the RMSE were used to compare simulated and measured water temperature, dissolved-oxygen concentration, selected nutrient concentrations, total organic carbon concentration, chlorophyll *a* concentration, and Secchi disk depth for the same three lake stations.

Certain measured water-quality values are censored, reported as "less than" values, and require special attention

before they can be included within statistical analyses. Censoring levels are specific to analytic methods for individual constituents and can change over time as methods change. The LRL for analytes measured at the USGS National Water Quality Laboratory generally is equal to twice the long-term method detection level (LT–MDL) and is intended to protect against false negatives. The LT–MDL is defined as the minimum concentration that can be measured, with 99-percent confidence, to be significantly greater than zero (Childress and others, 1999). One-half of the LT–MDL concentration was used for comparisons to simulated values. For instance, the LT–MDL for dissolved nitrate plus nitrite is 0.008 mg/L, and 0.004 mg/L was the value used to calculate mean concentrations for dissolved nitrate plus nitrite. Concentrations are marked as estimated for cases in which the concentration is between the LRL and the LT–MDL (Childress and others, 1999). Because there is a 95-percent confidence level that only 1 percent of concentrations above the LT-MDL are false positives, estimated concentrations were used as listed for statistical purposes in this report. MAE provides the error between the simulated and measured values and is computed by equation 1:

$$MAE = \sum \frac{\left|simulated\ value - measured\ value\right|}{number\ of\ observations} \quad (1)$$

For example, a MAE of 6.82°F means that the mean difference between simulated temperatures and measured temperatures is 6.82°F. The RMSE indicates the spread of how far simulated values deviate from the measured values and is computed by equation 2:

$$RMSE = \sqrt{\frac{\Sigma\left(simulated\ value - measured\ value\right)^2}{number\ of\ observations}} \quad (2)$$

For example, a RMSE of 8.25°F means that the simulated temperatures are within 8.25°F of the measured temperatures about 67 percent of the time. The NSE measures the magnitude of the differences between the measured and simulated values and is computed by equation 3:

$$NSE = 1 - \frac{\sum_{t=1}^{T}\left(measured\ value - simulated\ value\right)^2}{\sum_{t=1}^{T}\left(measured\ value - \overline{measured\ value}\right)^2} \quad (3)$$

where

T is time, in days; and

$\overline{measured\ value}$ is the average of all measured values.

A *NSE* of 1 indicates model predictions are a perfect match to the observed data.

Linear regression models developed by computer program S–LOADEST were used to estimate annual loads for each constituent for the HSPF model period at Williams Junction (table 5). Load is the mass of a constituent transported past a selected point in a stream in a given amount of time, in this case, 1 year. The S–LOADEST program (Runkel and others, 2004; TIBCO Software Inc., 2008) was used to estimate constituent loads by the rating-curve method (Cohn and others, 1989; Crawford, 1991). S–LOADEST estimates loads using mean daily streamflow, streamflow rating-curve parameters, several regression methods, and a ratio estimator. Because some of the constituent concentrations included in the S–LOADEST analyses were censored values, parameters were estimated by the adjusted maximum likelihood estimation (AMLE) method (Cohn, 1988; Cohn and others, 1992). In the absence of censored data, the method converts to the maximum likelihood estimation (MLE) method (Dempster and others, 1977; Wolynetz, 1979). Uncertainty in the estimated load was obtained using the method described by Likes (1980) and Gilroy and others (1990). The model (equation 4) used to calculate loads was based on the relation between the natural logarithms of L and Q:

$$ln(L) = b_0 + b_1\, ln(Q) \qquad (4)$$

where

ln	is natural logarithm;
L	is constituent load, in pounds per day;
b_0	is regression constant, dimensionless;
b_1	is a regression coefficient, dimensionless; and
Q	is daily mean streamflow, in cubic feet per second.

Estimated mean annual constituent loads and standard error of prediction (SEP) of the mean loads were calculated by S–LOADEST using all available data for each constituent for October 1989 through June 2011. The R^2 is the proportion of variability in the data set that is accounted for by the statistical model. Estimated residual variance is the MLE variance corrected for the number of observations, number of censored observations, and number of parameters in the regression model. Data from the Williams Junction water-quality and streamflow-gaging station generally appeared to fit the S–LOADEST models. Within the S–LOADEST model, the simple flow and concentration model was selected to eliminate storm-event (seasonal, in the case of the Lake Maumelle watershed) bias in the load estimations. This is because water-quality samples were not entirely random; they are specifically collected during storm events. Load regression models calculated by S–LOADEST had an estimated residual variance ranging from 0.385 (total organic carbon) to 3.055 (fecal coliform bacteria) and an R^2 ranging from 0.84 (dissolved orthophosphate) to 0.98 (total organic carbon) (table 5). The S–LOADEST regression models for total organic carbon, suspended sediment, and dissolved nitrite plus nitrate nitrogen were the best regression models (lowest variance, highest R^2). The fecal coliform bacteria and dissolved orthophosphorus models were the poorest regression models; however, all measured loads calculated with S–LOADEST were used in a comparison of flow-weighted concentrations from HSPF and S–LOADEST. The 95-percent confidence interval is the interval that has a 95-percent chance of containing the true regression line. A major factor determining the width of a

Table 5. Regression models developed using constituent concentrations from water samples collected at the Maumelle River at Williams Junction water-quality and streamflow-gaging station, October 1989 through June 2011.

[N, nitrogen; P, phosphorus; C, carbon; ln, natural logarithm; L, daily load in pounds per day, except fecal coliform load in million colonies per day; Q, daily mean streamflow in cubic feet per second]

Station name (number)	Constituent (parameter code)	Number of observations	Number of censored observations[1]	Regression model	Estimated residual variance[2]	Coefficient of determination (R^2)
Maumelle River at Williams Junction (07263295)	Dissolved ammonia, as N (00608)	136	59	ln(L) = -4.14 +0.97*lnQ	1.424	0.90
	Dissolved nitrite plus nitrate, as N (00631)	142	12	ln(L) = -3.64 + 1.07*lnQ	1.106	0.94
	Dissolved orthophosphate, as P (00671)	142	77	ln(L) = -6.50 + 1.00*lnQ	2.957	0.84
	Total phosphorus, as P (00665)	142	7	ln(L) = -3.68 + 1.00*lnQ	1.154	0.93
	Total organic carbon, as C (00680)	125	0	ln(L) = 1.37 + 1.04*lnQ	0.385	0.98
	Suspended sediment (80154)	139	0	ln(L) = 2.67 + 1.18*lnQ	0.885	0.95
	Fecal coliform (31625)	143	2	ln(L) = 5.83 + 1.15*lnQ	3.055	0.87

[1]Censored observations are a result of an analysis value lower than the laboratory minimum reporting level.

[2]Estimated residual variance is the maximum likelihood estimation variance corrected for the number of observations, number of censored observations, and number of parameters in the regression model.

confidence interval is the size of the sample used in the estimation procedure, with smaller samples having wider confidence intervals (Helsel and Hirsch, 2002). The confidence interval is included with the comparison of estimated loads calculated by S–LOADEST and simulated HSPF values to further qualify the accuracy of the regression models generated using measured values and S–LOADEST.

Loads were output directly from the HSPF model for Williams Junction and totaled on an annual basis. Results of the S–LOADEST loads then were compared to the HSPF simulated loads. Relative percentage difference (RPD) was used to evaluate the difference between the annual total S–LOADEST calculated loads and the HSPF simulated loads. The RPD was calculated using equation 5:

$$RPD = \left[|A - B| \middle/ \left(\frac{A+B}{2} \right) \right] \times 100 \qquad (5)$$

where

A and B are loads from each model.

Hydrologic Simulation Program– FORTRAN Model Calibration

Continuous streamflow data, discrete sediment concentration data, and other discrete water-quality data, in that sequential order, were used to calibrate the HSPF model. The model covers the period of January 1, 2003, through December 31, 2010, and uses the available stations with continuous streamflow gages and periodic water-quality sampling within the watershed for calibration (table 1). Calibration is a necessary step to verify the reliability of the model. Prior to calibration, the HSPF model was given meteorological data for the entire 2003 calendar year for the model to "warmup." This warmup period is a necessary step to allow for soil moisture to stabilize and constituent accumulation and washoff to equilibrate.

To calibrate water-quality loads entering streams, it is important to calibrate loads from each of the land-use types to known values. However, no studies have been conducted within the Lake Maumelle watershed for water-quality loads from the various land uses. The loading rate targets from each land use were obtained from literature values and can range substantially from study to study (Bureau of Land Management, 1983; U.S. Environmental Protection Agency, 1999; Maryland Department of the Environment, 2006; Scoles and others, 2001; Tetra Tech Inc., 2004). For the Lake Maumelle watershed, the lower end of the land-use loading range for water-quality loads was used for calibration (table 6). The lower end values

seemed to work well to calibrate the water-quality loads at each gage. The range for all parameter values used in the HSPF model streamflow calibration can be found in table 7.

Streamflow Calibration

Several HSPF model parameters (table 7) govern the simulated flux of water into a stream reach. Primary parameters that reflect soil conditions simulated from the PERLNDs (and therefore determine the distribution of available water for infiltration or for runoff) include: lower zone nominal storage (LZSN), upper zone nominal storage (UZSN), infiltration capacity of soil (INFILT), and lower zone evapotranspiration (LZETP). Parameters that affect surface runoff simulated from IMPLNDs are: length of the overland flow plane (LSUR), slope of the overland flow plane (SLSUR), Manning's n for the overland flow plane (NSUR), and the retention storage capacity of the surface (RETSC).

Streamflow calibration results (percent errors) for total flow volume and average daily mean flow rate (table 8) generally were within the acceptance rating criteria of "very good" to "good" (less than 10 to 15 percent) to "fair" (15 to 25 percent) for HSPF model performance (Donigian, 2000). In general, based on the exceedance probability, simulated "low flows" (in this instance, flows with exceedance probabilities greater than about 60 to 70 percent) were greater than the measured low flows for Bringle and Yount Creeks and Wye, but simulated high flows matched reasonably well to observed high flows (fig. 9). Streamflow calibration results were in close agreement at both high and low flows for Williams Junction (fig. 9). Simulated low flows were less than the measured low flows for Reece Creek, but simulated high flows matched reasonably well to observed high flows (fig. 9).

The simulated total flow volume and mean streamflow rates matched well with the measured data at all five stations used for the HSPF model calibration. Percent error between the measured and simulated total streamflow volumes at the five stations ranged from -10.24 to 15.73 percent (table 8). The percent error for Wye (the most downstream station and the largest inflow to Lake Maumelle) was only 6.57 percent. Williams Junction, upstream from Wye, simulated total streamflow volume was 9.06 percent lower than measured streamflow volume during the calibration period. Simulated total streamflow volumes for Wye and Yount were approximately 6 percent higher than measured streamflow volumes during the calibration period. Simulated streamflow volumes during the calibration period for Bringle were 15.73 percent higher than measured volumes and simulated volumes for Reece were 10.24 percent less than measured volumes.

Table 6. Land-use loading rates for the Hydrological Simulation Program—FORTRAN model of the Lake Maumelle watershed.

[Loading rates for constituents are sums of annual loads for the listed constituent divided by the acres for each land-use type; N, nitrogen; P, phosphorus; C, carbon; (ton/acre)/yr, ton per acre per year; (col/acre)/yr, colonies per acre per year; (lb/acre)/yr, pound per acre per year. Loading rate values are the low range of given literature values (Bureau of Land Management, 1983; Environmental Protection Agency, 1999; Maryland Department of the Environment, 2006; Scoles and others, 2001; Tetra Tech Inc., written commun., 2004)]

Land use	Suspended sediment ([ton/acre]/yr)	Fecal coliform bacteria ([col/acre]/yr)	Dissolved nitrite plus nitrate ([lb/acre]/yr as N)	Dissolved ammonia ([lb/acre]/yr as N)	Dissolved nitrite plus nitrate plus dissolved ammonia ([lb/acre]/yr as N)	Dissolved orthophosphate ([lb/acre]/yr as P)	Total phosphorus ([lb/acre]/yr as P)	Total organic carbon ([lb/acre]/yr as C)
Agriculture	0.228	1.30×10^{11}	0.114	0.050	0.164	0.067	0.319	23.4
Bare Soil	0.206	1.68×10^{9}	0.172	0.130	0.302	0.075	0.569	45.9
Clearcut	0.330	1.61×10^{9}	0.211	0.124	0.335	0.061	0.390	30.5
Coniferous	0.261	1.75×10^{9}	0.163	0.165	0.329	0.048	0.394	32.1
Deciduous	0.209	1.53×10^{9}	0.140	0.097	0.238	0.038	0.334	27.5
Grasslands	0.124	1.08×10^{11}	0.135	0.048	0.183	0.057	0.320	24.5
Paved Roads	0.303	6.73×10^{10}	0.185	0.122	0.308	0.052	0.495	41.1
Urban	0.462	6.86×10^{10}	0.166	0.118	0.284	0.090	0.519	39.9

Table 7. Summary of calibrated values for selected hydrology parameters for the Hydrological Simulation Program—FORTRAN model of the Lake Maumelle watershed.

[PERLND, pervious land area; IMPLND, impervious land area; numbers in bold are HSPF default values (Bicknell and others, 2001)]

Parameter	Land surface	Description	Values	Units
AGWETP	PERLND	Fraction of remaining evapotranspiration from active groundwater	0.00–0.05	dimensionless
AGWRC	PERLND	Base groundwater recession	0.85–0.95	1/day
BASETP	PERLND	Fraction of remaining evapotranspiration from base flow	0.025–0.060	dimensionless
CEPSC	PERLND	Interception storage capacity	0.03–0.20	inches
DEEPFR	PERLND	Fraction of groundwater inflow to deep recharge	0.211–0.251	dimensionless
INFEXP	PERLND	Infiltration equation exponent	**2.0**	dimensionless
INFILD	PERLND	Ratio of maximum and mean infiltration capacities	**2.0**	dimensionless
INFILT	PERLND	Index to infiltration capacity of soil	0.009–0.123	inches/interval
INTFW	PERLND	Interflow index	1.00–9.00	dimensionless
IRC	PERLND	Interflow recession coefficient	0.40–0.80	1/day
KVARY	PERLND	Groundwater outflow modifier	2.5	1/inches
LSUR	PERLND or IMPLND	Length of assumed overland flow plane	200–400	feet
LZETP	PERLND	Lower zone evapotranspiration	0.05–0.85	dimensionless
LZSN	PERLND	Lower zone nominal storage	4.0–10.2	inches
NSUR	PERLND or IMPLND	Manning's n for assumed overland flow plane	0.40	dimensionless
RETSC	IMPLND	Impervious retention storage capacity	0.10	inches
SLSUR	PERLND or IMPLND	Slope of assumed overland flow plane	0.011–0.309	dimensionless
UZSN	PERLND	Upper zone nominal storage	0.39–1.20	inches

Table 8. Hydrological Simulation Program—FORTRAN streamflow calibration results for the Lake Maumelle watershed.

[acre-ft, acre-feet; ft^3/s, cubic feet per second]

Maumelle River at Williams Junction (07263295)
Calibration period January 1, 2004 to December 31, 2010

Streamflow volumes[1]	Measured	Simulated	Percentage error
Total flow volume (acre-ft)	397,575	361,554	-9.06
Average daily mean flow rate (ft^3/s)	78.4	71.3	-9.06
Total of highest 10 percent of daily flows (acre-ft)	279,596	264,508	-5.40
Total of lowest 50 percent of daily flows (acre-ft)	8,158	8,210	0.64

Model-fit statistics[1]			
Number of days	2,557		
Coefficient of determination (R^2)	0.81		
Nash-Sutcliffe model efficiency coefficient	0.80		
Mean absolute error (ft^3/s)	47.41		
Root mean square error (ft^3/s)	124.44		

Maumelle River near Wye (07263296)
Calibration period July 11, 2007 to December 31, 2010

Streamflow volumes[1]	Measured	Simulated	Percentage error
Total flow volume (acre-ft)	337,660	359,847	6.57
Average daily mean flow rate (ft^3/s)	134.0	142.9	6.60
Total of highest 10 percent of daily flows (acre-ft)	242,757	256,219	5.55
Total of lowest 50 percent of daily flows (acre-ft)	1,836	2,170	18.22

Model-fit statistics[1]			
Number of days	1,270.00		
Coefficient of determination (R^2)	0.87		
Nash-Sutcliffe model efficiency coefficient	0.87		
Mean absolute error (ft^3/s)	68.99		
Root mean square error (ft^3/s)	186.14		

Bringle Creek at Martindale (072632962)
Calibration period May 7, 2005 to December 31, 2010

Streamflow volumes[1]	Measured	Simulated	Percentage error
Total flow volume (acre-ft)	43,796	50,687	15.73
Average daily mean flow rate (ft^3/s)	10.7	12.4	15.81
Total of highest 10 percent of daily flows (acre-ft)	27,724	32,869	18.56
Total of lowest 50 percent of daily flows (acre-ft)	1,233	3,532	186.41

Model-fit statistics[1]			
Number of days	2,065		
Coefficient of determination (R^2)	0.61		
Nash-Sutcliffe model efficiency coefficient	0.55		
Mean absolute error (ft^3/s)	7.83		
Root mean square error (ft^3/s)	23.84		

Table 8. Hydrological Simulation Program—FORTRAN streamflow calibration results for the Lake Maumelle watershed.—Continued

[acre-ft, acre-feet; ft³/s, cubic feet per second]

Yount Creek near Martindale (072632971)
Calibration period May 7, 2005 to December 31, 2010

Streamflow volumes[1]	Measured	Simulated	Percentage error
Total flow volume (acre-ft)	15,463	16,284	5.31
Average daily mean flow rate (ft³/s)	4.1	4.4	6.10
Total of highest 10 percent of daily flows (acre-ft)	11,730	11,071	-5.62
Total of lowest 50 percent of daily flows (acre-ft)	114	220	92.05

Model-fit statistics[1]			
Number of days	1,886		
Coefficient of determination (R²)	0.78		
Nash-Sutcliffe model efficiency coefficient	0.78		
Mean absolute error (ft³/s)	2.29		
Root mean square error (ft³/s)	5.74		

Reece Creek at Little Italy (072632982)
Calibration period May 7, 2005 to December 31, 2010

Streamflow volumes[1]	Measured	Simulated	Percentage error
Total flow volume (acre-ft)	34,297	30,785	-10.24
Average daily mean flow rate (ft³/s)	10.6	9.6	-9.91
Total of highest 10 percent of daily flows (acre-ft)	22,276	20,144	9.57
Total of lowest 50 percent of daily flows (acre-ft)	1,039	752	-27.66

Model-fit statistics[1]			
Number of days	1,626		
Coefficient of determination (R²)	0.75		
Nash-Sutcliffe model efficiency coefficient	0.73		
Mean absolute error (ft³/s)	5.99		
Root mean square error (ft³/s)	18.62		

[1]Only paired data were used in statistical calculations; missing data were not used.

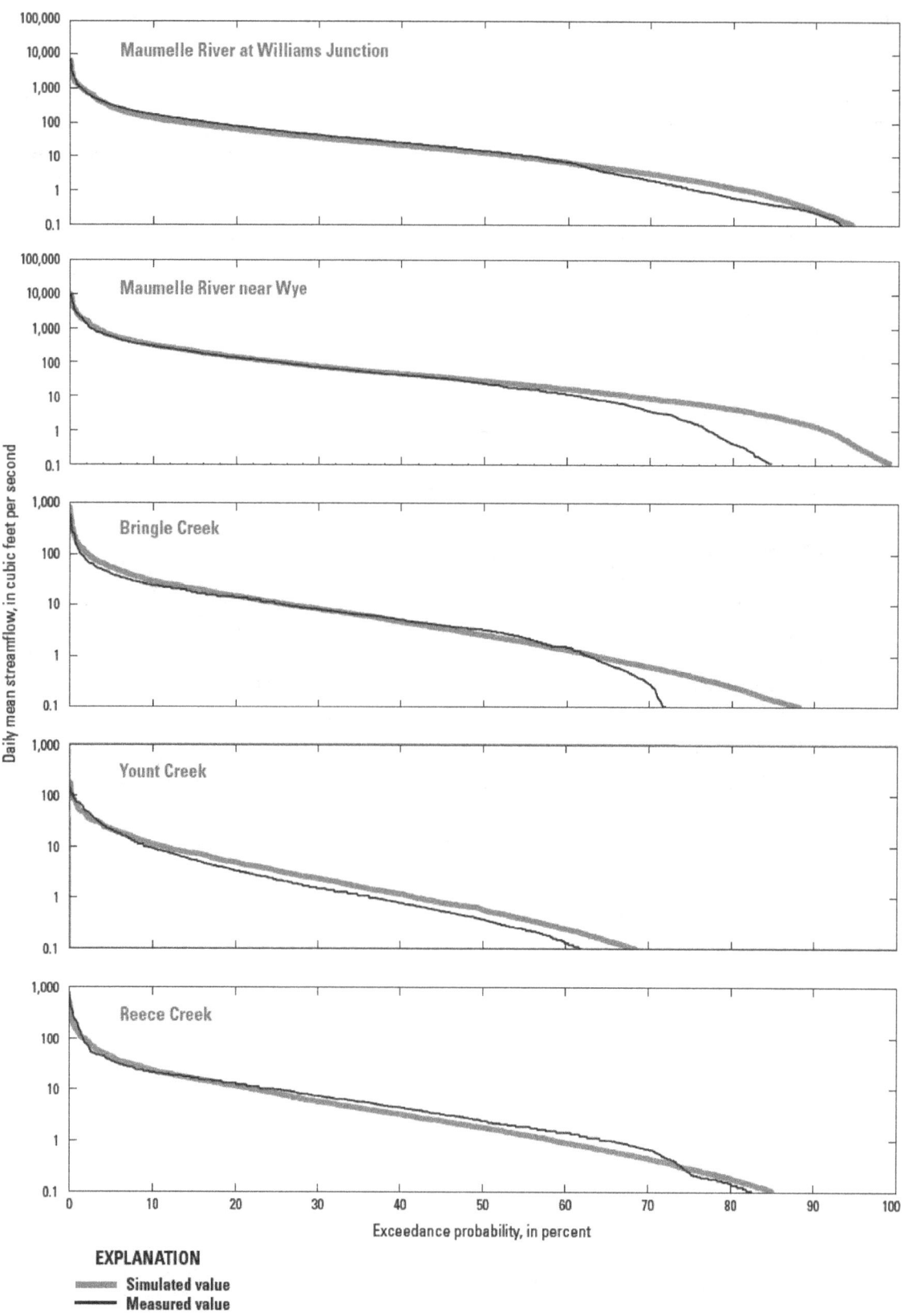

Figure 9.   Streamflow exceedance probability of simulated and measured streamflow at selected inflow stations in the Lake Maumelle watershed.

Temporally, the streamflow calibration matched the seasonal variability occurring during base-flow conditions (streamflows not substantially influenced by runoff) and during high flows (fig. 10). Measured and simulated values matched closely for the total of highest 10 percent of daily flows and total of lowest 50 percent of daily flows for the two stations on Maumelle River, Williams Junction and Wye (fig. 10; table 8). The majority of streamflow entering Lake Maumelle is contributed by Maumelle River, thus calibration at the two Maumelle River stations was necessary for the water balance (base flow and high flows) to be accurate within the watershed model and reservoir model.

Sediment Calibration

Two equations are used in HSPF to calculate the production, removal, and transport of sediment from the land surface for three sediment size classes (sand, silt, and clay). These equations are included in the Agricultural Runoff Model (ARM) and Nonpoint Source (NPS) models developed by Donigian and Davis (1978) and Donigian and Crawford (1977), respectively. The production of sediment is simulated by detachment of soil by precipitation, the removal of sediment by scour of the soil matrix, and transport of the detached sediment by overland flow. Sediment load within a stream reach is calculated from particle size, soil texture, settling velocity, density, and erodibility and is simulated by convection, scouring, and deposition. Suspended-sediment concentrations were measured for all inflow stations according to the sample interval in table 1. Value ranges and units for calibrated HSPF water-quality parameters, including those used in the simulation of suspended sediment, are listed in table 9.

In general, simulated and measured suspended-sediment concentrations during periods of base flow (streamflows not substantially influenced by runoff) agreed reasonably well for Williams Junction (fig. 11), with difference (simulated minus measured value, 80 percent of the values) ranging from -15 to 41 mg/L, and percent difference—relative to the measured value—generally ranging from -99 to 182 percent and Wye (differences generally ranging from -20 to 22 mg/L, -100 to 194 percent). Additionally, simulated suspended-sediment concentrations matched well with the quarterly and monthly sampling values and also during periods of stormflow (streamflow substantially influenced by runoff) for all stations (fig. 11). For paired (measured values and mean of simulated daily values for days with measured values) suspended-sediment concentrations at Williams Junction, the mean measured suspended-sediment concentration was 25 mg/L and the mean simulated daily suspended-sediment concentration was 27 mg/L (table 10). The RPDs between the annual S–LOADEST and simulated HSPF suspended-sediment loads ranged from 24.44 to 84.25 percent with a median value of 51.40 percent (table 11).

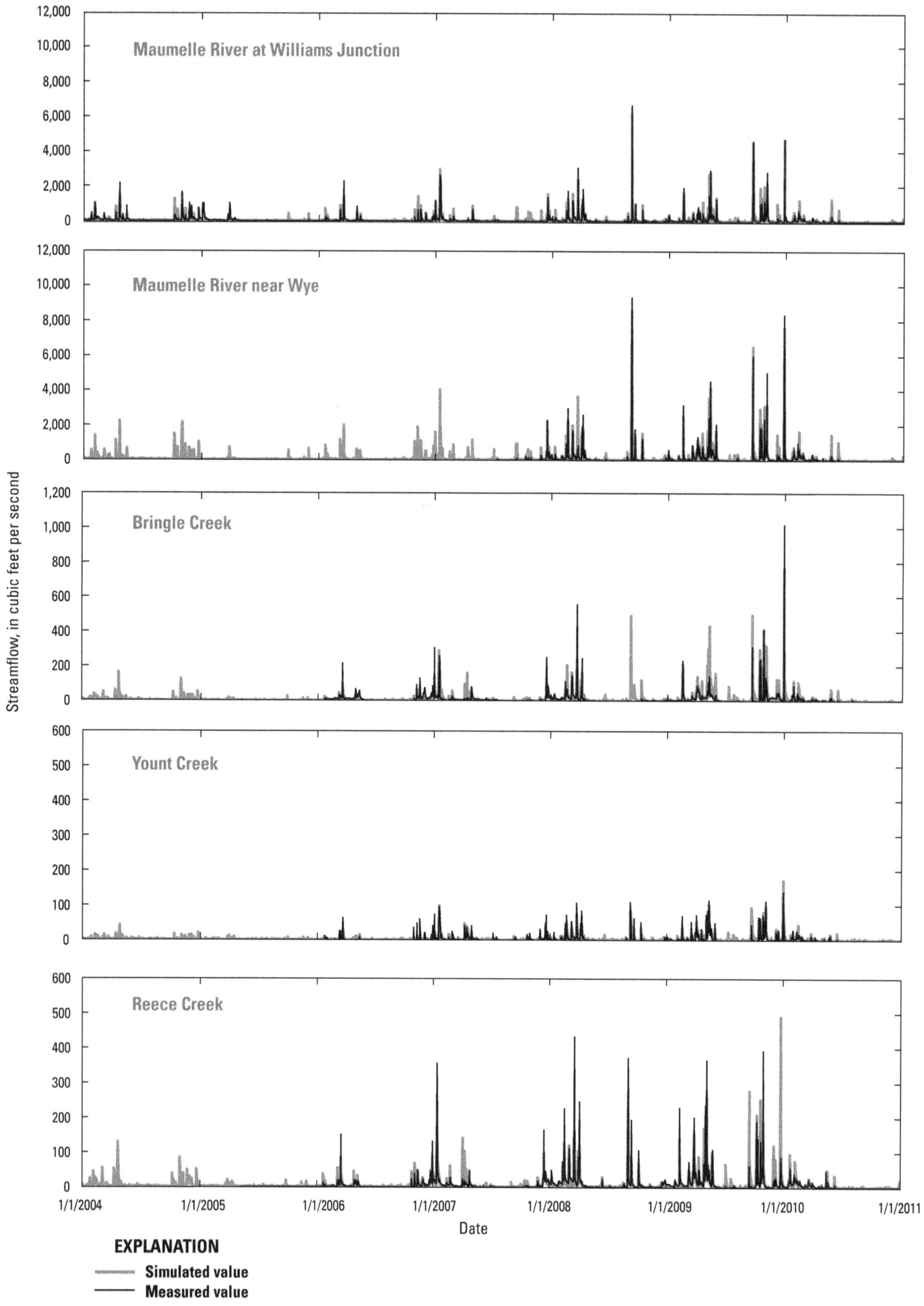

Figure 10. Simulated and measured streamflow at selected inflow stations in the Lake Maumelle watershed.

Table 9.   Summary of calibrated values for selected water-quality parameters for Hydrological Simulation Program—FORTRAN model   f of the Lake Maumelle watershed.

[RCHRES, stream reach; PERLND, pervious land surface; IMPLND, impervious land surface; GQUAL, general quality constituent; --, none; #, number; numbers in bold are HSPF default values (Bicknell and others, 2001)]

Parameter	Model unit	Description	Values	Units
		Water temperature		
CFSAEX	RCHRES	Correction factor for solar radiation, the fraction of the RCHRES surface exposed to solar radiation	0.90–1.80	--
KATRAD	RCHRES	Longwave radiation coefficient	**9.37**–17.37	--
KCOND	RCHRES	Conduction-convection heat transport coefficient	12.12–15.12	--
KEVAP	RCHRES	Evaporation coefficient	**2.24**	--
		Dissolved oxygen		
IDOXP	PERLND	Concentration of dissolved oxygen in interflow	8.8	milligrams per liter
ADOXP	PERLND	Concentration of dissolved oxygen in base flow	8.8	milligrams per liter
KBOD20	RCHRES	Unit biochemical oxygen demand decay rate at 20 degrees Celsius	0.01	/hour
KODSET	RCHRES	Rate of biochemical oxygen demand settling	0.127	feet/hour
SUPSAT	RCHRES	Maximum allowable dissolved oxygen supersaturation factor	**1.15**	--
BENOD	RCHRES	Benthal oxygen demand at 20 degrees Celsius	50	milligram/square meter-hour
BRBOD(1)	RCHRES	Benthal release rate of biochemical oxygen demand under anaerobic conditions	1	milligram/square meter-hour
BRBOD(2)	RCHRES	Increment to benthal release rate of biochemical oxygen demand under anaerobic conditions	1	milligram/square meter-hour
EXPREL	RCHRES	Exponent in the dissolved oxygen term of the benthal biochemical oxygen demand release equation	**2.82**	--
		Suspended sediment		
KRER	PERLND	Coefficient of the soil-detachment equation	0.10–0.74	complex
JRER	PERLND	Exponent of the soil-detachment equation	2.0–2.1	complex
KSER	PERLND	Coefficient of the detached-sediment washoff equation	0.030–4.175	complex
JSER	PERLND	Exponent of the detached-sediment washoff equation	1.50–1.70	complex
AFFIX	PERLND	Fraction by which detached sediment decreases daily through soil compaction	0.05–0.10	1/day
COVER	PERLND	Fraction of the land surface shielded from rainfall erosion	0.50–0.95	--
NVSI	PERLND	Rate at which sediment enters detached storage from the atmosphere	**0**	pound/acre-day
KEIM	IMPLND	Coefficient of the solids washoff equation	0.30–0.70	complex
JEIM	IMPLND	Exponent of the solids washoff equation	2.00–2.50	complex
ACCSDP	IMPLND	Solids accumulation rate	0.0005	ton/acre-day
RHO	RCHRES	Density of the sediment particle	2.0–2.3	gram/cubic centimeter
M (silt)1	RCHRES	Erodibility coefficient of the sediment	0.001–0.01	pound/square foot-hour
M (clay)1	RCHRES	Erodibility coefficient of the sediment	0.0001–0.001	pound/square foot-hour
W (silt and clay)	RCHRES	Settling velocity of the sediment particle in still water	0.0005–0.005	inch/second
TAUCD (silt)	RCHRES	Critical bed shear stress for sediment deposition	0.001–0.3	pound/square foot
TAUCS (silt)	RCHRES	Critical bed shear stress for sediment scour	0.012–2.4	pound/square foot
TAUCD (clay)	RCHRES	Critical bed shear stress for sediment deposition	0.001–0.3	pound/square foot
TAUCS (clay)	RCHRES	Critical bed shear stress for sediment scour	0.012–2.4	pound/square foot

Table 9.   Summary of calibrated values for selected water-quality parameters for Hydrological Simulation Program—FORTRAN model of the Lake Maumelle watershed.—Continued

[RCHRES, stream reach, PERLND, pervious land surface, IMPLND, impervious land surface, GQUAL, general quality constituent; --, none; #, number; numbers in bold are HSPF default values (Bicknell and others, 2001)]

Parameter	Model unit	Description	Values	Units
		Dissolved ammonia nitrogen (GQUAL)		
ACQOP	PERLND	Accumulation rate of constituent on surface	0.003–3.0	pound/acre-day
SQOLIM	PERLND	Maximum storage of constituent on surface	0.003–1.0	pound/acre
WSQOP	PERLND	Rate of surface runoff to remove 90 percent of stored constituent in 1 hour	1.0–1.5	inch/hour
KTAM20	RCHRES	Nitrification rate of ammonia at 20 degrees Celsius	0.05–0.08	/hour
KNO220	RCHRES	Nitrification rate of nitrite at 20 degrees Celsius	0.035	/hour
		Dissolved nitrate nitrogen (GQUAL)		
ACQOP	PERLND	Accumulation rate of constituent on surface	0.00001–0.1	pound/acre-day
SQOLIM	PERLND	Maximum storage of constituent on surface	0.0001–1.0	pound/acre
WSQOP	PERLND	Rate of surface runoff to remove 90 percent of stored constituent in 1 hour	1.0–1.5	inch/hour
KNO320	RCHRES	Denitrification rate of nitrate at 20 degrees Celsius	0.50	/hour
		Dissolved orthophosphate (GQUAL)		
WSQOP	PERLND	Rate of surface runoff to remove 90 percent of stored constituent in 1 hour	0.50–0.80	inch/hour
POTFW	PERLND	Potency factor of sediment in washoff	0.50–4.33	pound/ton
POTFS	PERLND	Scour potency factor of sediment in washoff	0.00	pound/ton
		Fecal coliform bacteria (GQUAL)		
ACCUM	PERLND	Accumulation rate of constituent on surface	6×10^6–1×10^{11}	# organisms/acre-day
SQOLIM	PERLND	Maximum storage of constituent on surface	9×10^8–2×10^{11}	# organisms/acre
WSQOP	PERLND	Rate of surface runoff to remove 90 percent of stored constituent in 1 hour	0.5	inch/hour
FSTDEC	RCHRES	First-order decay rate for quality constituent	0.24	# organisms/day
		Total organic carbon (GQUAL)		
ACQOP	PERLND	Accumulation rate of constituent on surface	1.20	pounds/acre-day
SQOLIM	PERLND	Maximum storage of constituent on surface	5.10–75.10	pounds/acre
WSQOP	PERLND	Rate of surface runoff to remove 90 percent of stored constituent in 1 hour	0.5–1.0	inch/hour

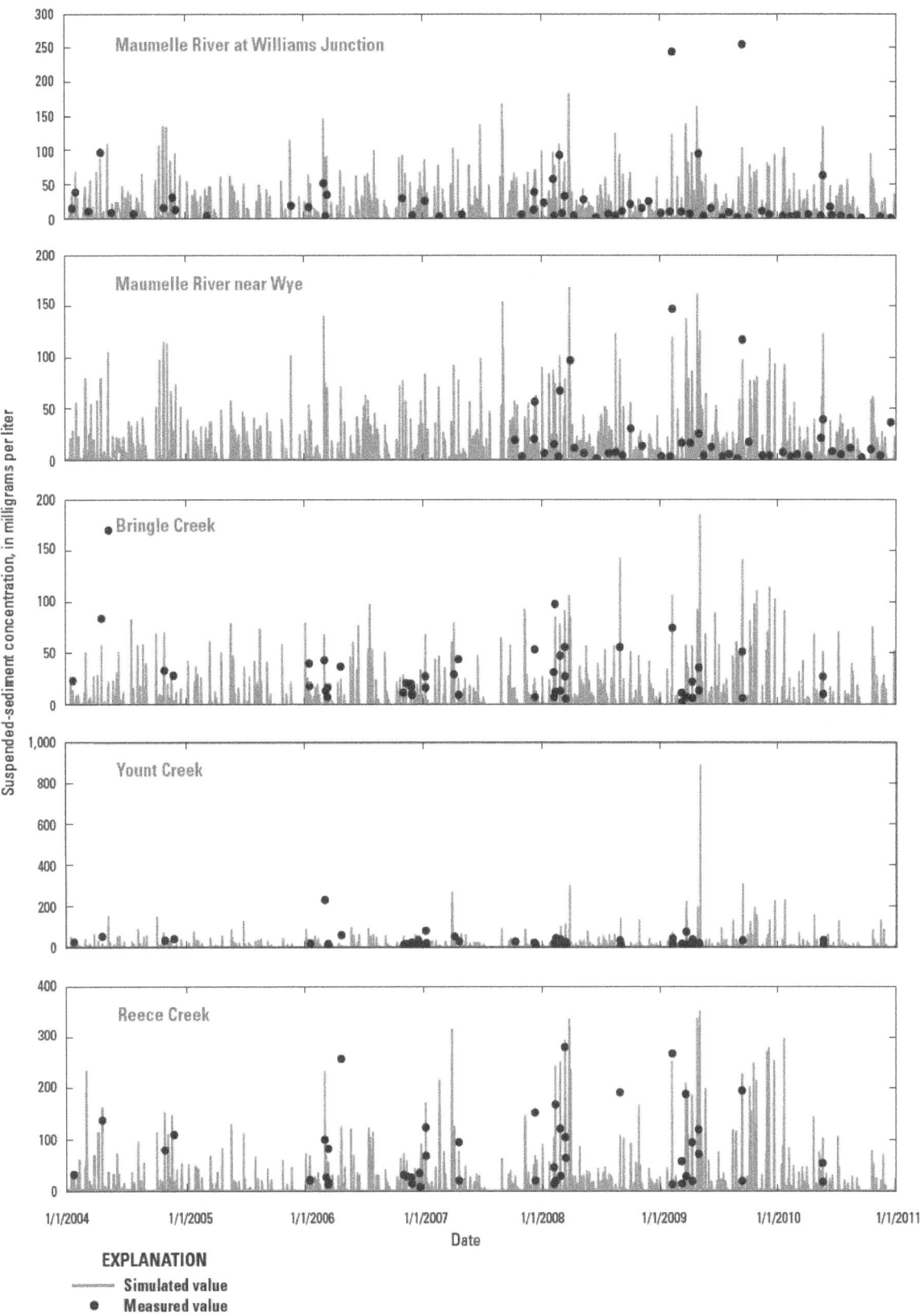

Figure 11. Simulated and measured mean daily suspended-sediment concentrations at selected inflow stations in the Lake Maumelle watershed.

Table 10. Hydrological Simulation Program—FORTRAN model calibration evaluation statistics for Maumelle River at Williams Junction.

[mg/L, milligrams per liter; F, degrees Fahrenheit; col/100mL, colonies per 100 milliliters; P, phosphorus; <, less than; N/A, not applicable; N, nitrogen; C, carbon; LRL, laboratory reporting level; the measured instantaneous concentrations, collected from January 1, 2004, through December 31, 2010, were averaged. Hourly simulated values were aggregated to daily values and paired with measured dates then averaged]

Water-quality characteristic	Paired data for days with measured data		Number of days	Difference between means of paired data (percent)
	Mean of measured data	Mean of simulated daily values		
Maumelle River at Williams Junction (07263295) Calibration period January 1, 2004 to December 31, 2010				
Suspended sediment (mg/L)[1]	25	27	66	9
Water temperature (°F)	58.4	53.9	66	-8
Dissolved oxygen (mg/L)	9.0	10.5	66	16
Fecal coliform bacteria (col/100 mL)[2]	389	105	66	-73
Dissolved orthophosphate (mg/L as P)[3]	<0.008	0.003	65	N/A[8]
Total phosphorus (mg/L as P)[4]	0.032	0.043	65	34
Dissolved nitrite plus nitrate (mg/L as N)[5]	0.059	0.030	65	-49
Dissolved ammonia (mg/L as N)[6]	<0.02	0.017	65	N/A[8]
Total organic carbon (mg/L as C)[7]	5.3	4.1	65	-23

[1]Suspended sediment LRL: 1 mg/L.

[2]Fecal coliform LRL: not applicable.

[3]Dissolved orthophosphate LRL: 0.006 mg/L; after October 1, 2008, LRL: 0.008 mg/L.

[4]Total phosphorus LRL: 0.004 mg/L; after October 1, 2008, LRL: 0.008 mg/L.

[5]Dissolved nitrite plus nitrate LRL: 0.016 mg/L; after October 1, 2010, LRL: 0.008 mg/L.

[6]Dissolved ammonia LRL: 0.01 mg/L; after October 1, 2006, LRL: 0.02 mg/L.

[7]Total organic carbon LRL: 0.40 mg/L; after October 1, 2008, LRL: 0.60 mg/L.

[8]Percentage difference was not calculated when mean of measured values was equal or less than laboratory reporting level.

Table 11. Annual loads from the S-LOADEST and Hydrologic Simulation Program–FORTRAN models for Maumelle River at Williams Junction.

[N, nitrogen; P, phosphorus; C, carbon; %, percent; lbs, pounds; col/100mL, colonies per 100 milliliters; LRL, laboratory reporting level]

Constituent	S-LOADEST			HSPF	Relative percentage difference
Dissolved ammonia as N[1]	Lower 95% load (lbs)[2]	Total load (lbs)	Upper 95% load (lbs)[3]	Total load (lbs)	
2004	1,886.48	1,889.46	1,897.81	3,729.71	65.50
2005	698.91	699.82	703.39	826.15	16.56
2006	1,068.99	1,070.45	1,075.73	3,097.85	97.28
2007	1,244.00	1,245.57	1,252.08	3,333.27	91.19
2008	2,260.70	2,263.53	2,275.35	5,122.24	77.41
2009	3,093.34	3,097.77	3,112.56	9,736.24	103.45
2010	693.31	694.44	697.44	2,212.12	104.43

Constituent	S-LOADEST			HSPF	Relative percentage difference
Dissolved nitrite plus nitrate as N[4]	Lower 95% load (lbs)[2]	Total load (lbs)	Upper 95% load (lbs)[3]	Total load (lbs)	
2004	8,888.99	8,904.06	8,941.30	5,508.67	47.12
2005	3,201.87	3,206.43	3,221.80	1,135.79	95.37
2006	5,011.52	5,018.89	5,042.39	4,331.05	14.71
2007	5,937.77	5,945.68	5,975.57	5,339.43	10.74
2008	11,405.19	11,420.15	11,477.92	7,285.92	44.20
2009	15,739.71	15,763.80	15,835.45	10,536.30	39.75
2010	2,962.68	2,967.94	2,979.97	2,968.09	0.01

Constituent	S-LOADEST			HSPF	Relative percentage difference
Dissolved orthophosphate as P[5]	Lower 95% load (lbs)[2]	Total load (lbs)	Upper 95% load (lbs)[3]	Total load (lbs)	
2004	782.19	782.78	788.34	352.04	75.91
2005	286.86	287.00	289.57	77.82	114.68
2006	442.10	442.34	446.12	292.40	40.82
2007	516.77	517.00	521.83	332.61	43.41
2008	955.70	956.12	965.04	418.20	78.28
2009	1,312.94	1,313.73	1,324.33	601.41	74.39
2010	278.64	278.87	280.77	196.58	34.62

Constituent	S-LOADEST			HSPF	Relative percentage difference
Total phosphorus as P[6]	Lower 95% load (lbs)[2]	Total load (lbs)	Upper 95% load (lbs)[3]	Total load (lbs)	
2004	5,574.60	5,584.15	5,607.30	9,154.41	48.45
2005	2,042.39	2,045.35	2,055.03	1,637.90	22.12
2006	3,151.05	3,155.79	3,170.32	7,341.00	79.74
2007	3,688.49	3,693.61	3,711.64	7,966.94	73.30
2008	6,843.03	6,852.47	6,885.94	10,064.95	37.98
2009	9,401.32	9,416.07	9,458.09	14,680.37	43.69
2010	1,976.59	1,980.10	1,988.12	4,965.42	85.96

Table 11. Annual loads from the S-LOADEST and Hydrologic Simulation Program–FORTRAN models for Maumelle River at Williams Junction.—Continued

[N, nitrogen; P, phosphorus; C, carbon; %, percent; lbs, pounds; col/100mL, colonies per 100 milliliters; LRL, laboratory reporting level]

Constituent	S-LOADEST			HSPF	Relative percentage difference
Total organic carbon as C[7]	Lower 95% load (lbs)[2]	Total load (lbs)	Upper 95% load (lbs)[3]	Total load (lbs)	
2004	1,032,127	1,034,374	1,037,883	859,742	18.44
2005	374,629	375,385	376,752	155,120	83.04
2006	582,462	583,656	585,754	691,993	16.99
2007	686,048	687,397	689,956	746,144	8.20
2008	1,296,743	1,299,274	1,304,119	943,400	31.74
2009	1,786,394	1,790,119	1,796,459	1,373,302	26.35
2010	354,181	354,968	356,156	474,416	28.80

Constituent	S-LOADEST			HSPF	Relative percentage difference
Suspended sediment[8]	Lower 95% load (lbs)[2]	Total load (lbs)	Upper 95% load (lbs)[3]	Total load (lbs)	
2004	5,204,001	5,213,131	5,234,347	6,664,296	24.44
2005	1,827,996	1,830,729	1,839,200	745,514	84.25
2006	2,933,523	2,937,981	2,951,409	4,748,064	47.10
2007	3,577,639	3,582,500	3,600,265	6,232,742	54.00
2008	7,305,052	7,314,655	7,351,612	11,612,997	45.42
2009	10,088,461	10,104,315	10,149,343	17,341,035	52.74
2010	1,549,659	1,552,539	1,558,600	2,626,658	51.40

Constituent	S-LOADEST			HSPF	Relative percentage difference
Fecal coliform bacteria[9]	Lower 95% load (col/100mL)[2]	Total load (col/100mL)	Upper 95% load (col/100mL)[3]	Total load (col/100mL)	
2004	5.15×10^{13}	5.15×10^{13}	5.19×10^{13}	1.09×10^{14}	71.66
2005	1.82×10^{13}	1.82×10^{13}	1.84×10^{13}	3.38×10^{13}	59.90
2006	2.90×10^{13}	2.90×10^{13}	2.93×10^{13}	8.57×10^{13}	98.86
2007	3.50×10^{13}	3.50×10^{13}	3.54×10^{13}	9.20×10^{13}	89.80
2008	7.00×10^{13}	7.00×10^{13}	7.08×10^{13}	1.48×10^{14}	71.56
2009	9.68×10^{13}	9.68×10^{13}	9.77×10^{13}	2.24×10^{14}	79.30
2010	1.59×10^{13}	1.59×10^{13}	1.60×10^{13}	6.61×10^{13}	122.39

[1]Dissolved ammonia LRL: 0.01 mg/L; after October 1, 2006, LRL: 0.02 mg/L.

[8]Lower bound of 95 percent confidence interval of flow-weighted concentrations.

[9]Upper bound of 95 percent confidence interval of flow-weighted concentrations.

[2]Dissolved nitrite plus nitrate LRL: 0.016 mg/L.

[3]Dissolved orthophosphate LRL: 0.006 mg/L; after October 1, 2008, LRL: 0.008 mg/L.

[4]Total phosphorus LRL: 0.004 mg/L; after October 1, 2008, LRL: 0.008 mg/L.

[5]Total organic carbon LRL: 0.40 mg/L; after October 1, 2008, LRL: 0.60 mg/L.

[6]Suspended sediment LRL: 1 mg/L.

[7]Fecal coliform LRL: not applicable.

The median (2004 through 2010) loading rates for suspended sediment for subwatersheds draining directly into Lake Maumelle (the rates were calculated from the area upstream from each of these subwatersheds plus the area in the subwatershed) ranged from 0.05 to 0.17 (tons/acre)/yr (table 12). The median suspended-sediment loading rate for subwatershed 22 (the area upstream from the mouth of the Maumelle River; the largest subwatershed in the Lake Maumelle watershed) was 0.11 (ton/acre)/yr.

Water Temperature Calibration

Water temperature calibration is an important step for proper calibration of the processes that influence water quality within the stream reaches. Water temperature within the HSPF model is simulated by three main processes: (1) the exchange of heat through water into and out of the reach, (2) heat exchange between the streambed and the reach, and (3) heat exchange between the air and water surface. The exchange of heat at the air-water interface had the greatest effect on water temperature simulation because of the observed sensitivity of the model to the atmospheric longwave radiation coefficient (KATRAD). Water temperature measurements were available for all stream reaches according to the sample period in table 1.

Water temperature simulations followed measured water temperature seasonal trends for all stations with the largest differences occurring during approximately December through February (fig. 12). The differences—simulated minus measured value—generally ranged from -11°F to 22°F for Williams Junction, and Wye differences generally ranged from -23°F to 10°F. For paired (measured values and simulated mean daily values for days with measured values) water temperatures at Williams Junction, the mean measured water temperature was 58.4°F and the mean daily simulated water temperature was 53.9°F (table 10).

Dissolved-Oxygen Calibration

Quantifying dissolved oxygen is important in the modeling of nutrients and is an important indicator of overall aquatic ecosystem health. Oxygen can enter a stream reach through incoming tributaries, from PERLNDs or IMPLNDs through the same processes that water enters a reach, and from reaeration from the atmosphere. Dissolved-oxygen losses include outflow at the downstream end of the reach, nitrification, and benthal oxygen demand; whereas, growth and respiration of phytoplankton and benthic organisms can cause a loss or gain of dissolved oxygen. Dissolved-oxygen concentrations were measured at inflow and lake stations according to the sample period in table 1.

Dissolved-oxygen simulations followed measured dissolved-oxygen seasonal trends for all stations with the largest differences occurring during approximately July through August (during the periods of lowest measured dissolved-oxygen concentrations); in other months, the simulated dissolved-oxygen concentrations generally were less than 1 mg/L less than measured dissolved-oxygen concentrations (fig. 13). For paired (measured values and simulated daily mean values for days with measured values) dissolved-oxygen concentrations at Williams Junction, the mean measured dissolved-oxygen concentration was 9.0 mg/L and the mean daily simulated dissolved-oxygen concentration was 10.5 mg/L (table 10).

Fecal Coliform Bacteria Calibration

Fecal coliform bacteria concentration is simulated within HSPF as a general quality constituent as the number of organisms on the PERLND and IMPLND surface that can be input into a stream reach by direct runoff or attached to sediment. Monthly accumulation and monthly fecal coliform bacteria storage values were calculated using the Bacterial Indicator Tool developed by the U.S. Environmental Protection Agency (2000). The purpose of the Bacterial Indicator Tool is to estimate fecal coliform bacteria contributions from multiple sources including livestock, poultry, and septic systems. Agricultural census data (U.S. Department of Agriculture, 2007) for each county in the study area were used to calculate the number of agricultural animals for each land-use type within the watershed. The mean fecal coliform bacteria numbers calculated from the Bacterial Indicator Tool for each month were input into the monthly accumulation and monthly storage tables of the HSPF model as number of organisms per acre per day and number of organisms per acre, respectively. Fecal coliform bacteria in groundwater and interflow were not considered in the HSPF model.

Table 12. Baseline condition simulated subwatershed water-quality median loading rates and mean streamflows for subwatershed immediately adjacent to Lake Maumelle watershed, Arkansas, from the Hydrological Simulation Program—FORTRAN model.

[Loading rates for subwatersheds are sums of annual loads for the listed subwatershed and upstream subwatersheds divided by the acres in these subwatersheds; N, nitrogen; P, phosphorus; C, carbon; (lb/acre)/yr, pound per acre per year; (ton/acre)/yr, ton per acre per year; (col/acre)/yr, colonies per acre per year; ft³/s, cubic foot per second; mean values calculated using years 2004–10]

Subwatershed	Dissolved nitrite plus nitrate ([lb/acre]/yr as N)	Dissolved ammonia ([lb/acre]/yr as N)	Dissolved orthophosphate ([lb/acre]/yr as P)	Total phosphorus ([lb/acre]/yr as P)	Total organic carbon ([lb/acre]/yr as C)	Suspended sediment ([ton/acre]/yr)	Fecal coliform bacteria ([col/acre]/yr)	Mean streamflow (ft³/s)
15 (Maumelle River at Williams Junction; Upper watershed area)[1]	0.182	0.138	0.011	0.273	25.7	0.12	3.72×10^9	71.28
18 (Maumelle River near Wye; Critical area B)[1]	0.171	0.127	0.033	0.346	30.0	0.11	5.96×10^9	108.1
20 (Bringle Creek; Critical area B)[1]	0.103	0.059	0.020	0.139	11.7	0.05	4.78×10^9	11.38
22 (Maumelle River, Critical area B)	0.178	0.130	0.037	0.355	30.5	0.11	6.18×10^9	123.5
31 (Yount Creek; Critical area B)	0.140	0.039	0.016	0.225	18.3	0.05	7.89×10^9	4.37
25 (Reece Creek; Critical area B)	0.133	0.078	0.031	0.240	18.4	0.16	1.04×10^{10}	11.12
26 (Critical area B)	0.118	0.090	0.013	0.213	15.9	0.16	1.31×10^9	0.62
28 (Critical area B)	0.138	0.089	0.018	0.259	19.7	0.16	2.87×10^9	3.80
29 (Critical area B)	0.081	0.075	0.003	0.189	14.3	0.14	1.01×10^{10}	0.22
33 (Critical area A)	0.130	0.094	0.014	0.231	17.6	0.16	8.30×10^9	0.91
35 (Critical area A)	0.127	0.090	0.010	0.242	18.2	0.17	1.79×10^9	1.32
37 (Critical area B)	0.135	0.088	0.018	0.247	18.7	0.16	5.70×10^9	4.42
39 (Critical area A)	0.139	0.102	0.012	0.248	18.9	0.16	3.79×10^9	1.26
41 (Critical area A)	0.139	0.098	0.012	0.248	18.8	0.17	1.66×10^9	1.63
43 (Critical area A)	0.141	0.095	0.018	0.254	19.3	0.16	1.60×10^{10}	2.76
45 (Critical area B)	0.140	0.095	0.023	0.248	19.4	0.13	7.60×10^9	2.28
47 (Critical area A)	0.136	0.097	0.017	0.246	18.8	0.16	1.06×10^{10}	0.89
49 (Critical area B)	0.141	0.086	0.025	0.267	20.6	0.17	2.37×10^9	4.48
51 (Critical area B)	0.136	0.088	0.019	0.255	19.6	0.15	4.11×10^9	3.78
53 (Critical area B)	0.141	0.090	0.019	0.273	20.9	0.17	3.46×10^9	4.24
Loading rate and streamflow for entire watershed (baseline)[2,3]	0.164	0.114	0.032	0.320	26.7	0.12	6.34×10^9	8.58

[1]Components (upstream subwatersheds) of subwatershed 22.

[2]To summarize the loading rate for the entire watershed, the constituent loads for all subwatersheds were summed together, then divided by the total acres present within the entire watershed.

[3]Watershed does not include Lake Maumelle surface.

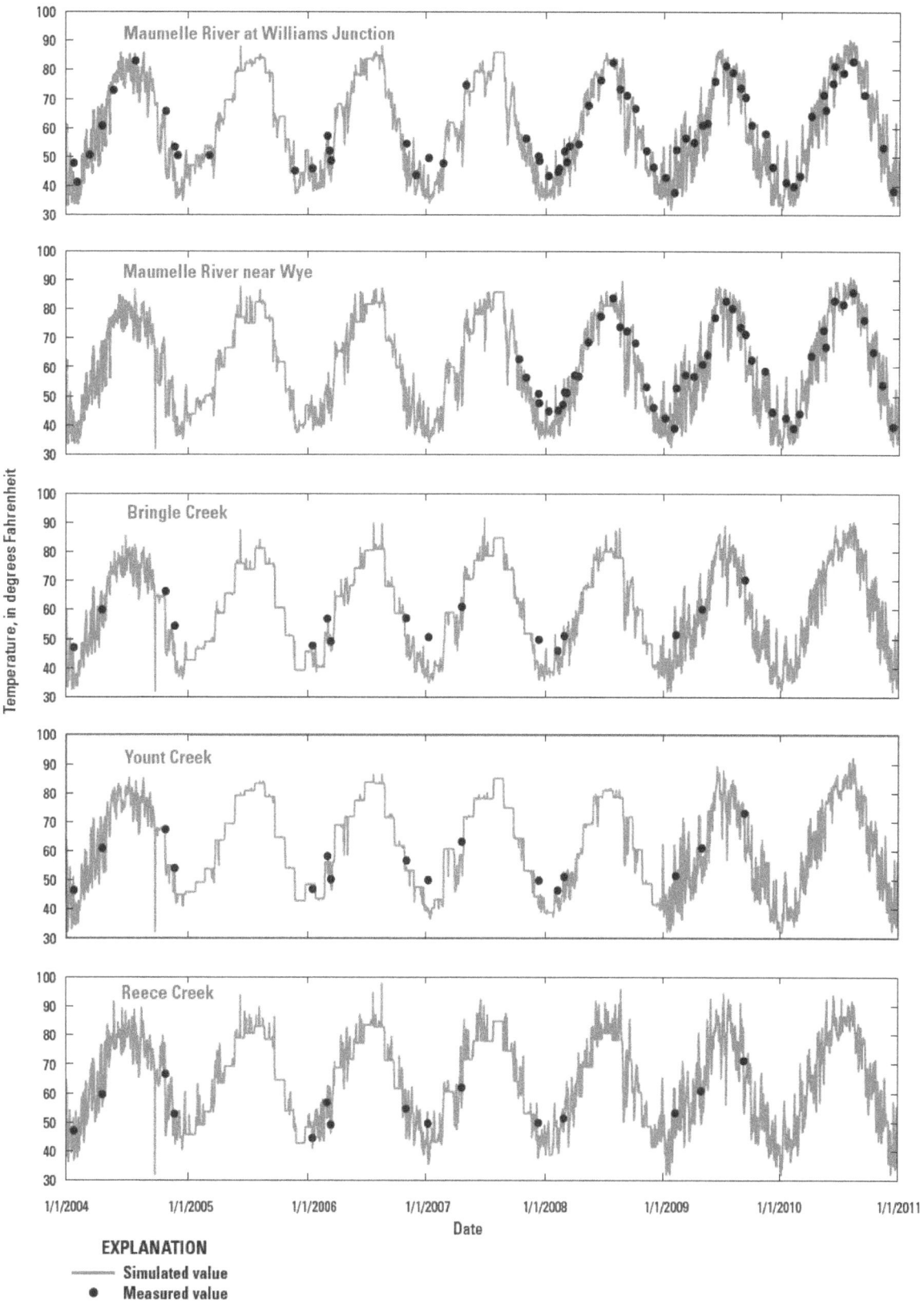

Figure 12.   Simulated and measured mean daily temperature at inflow stations in the Lake Maumelle watershed.

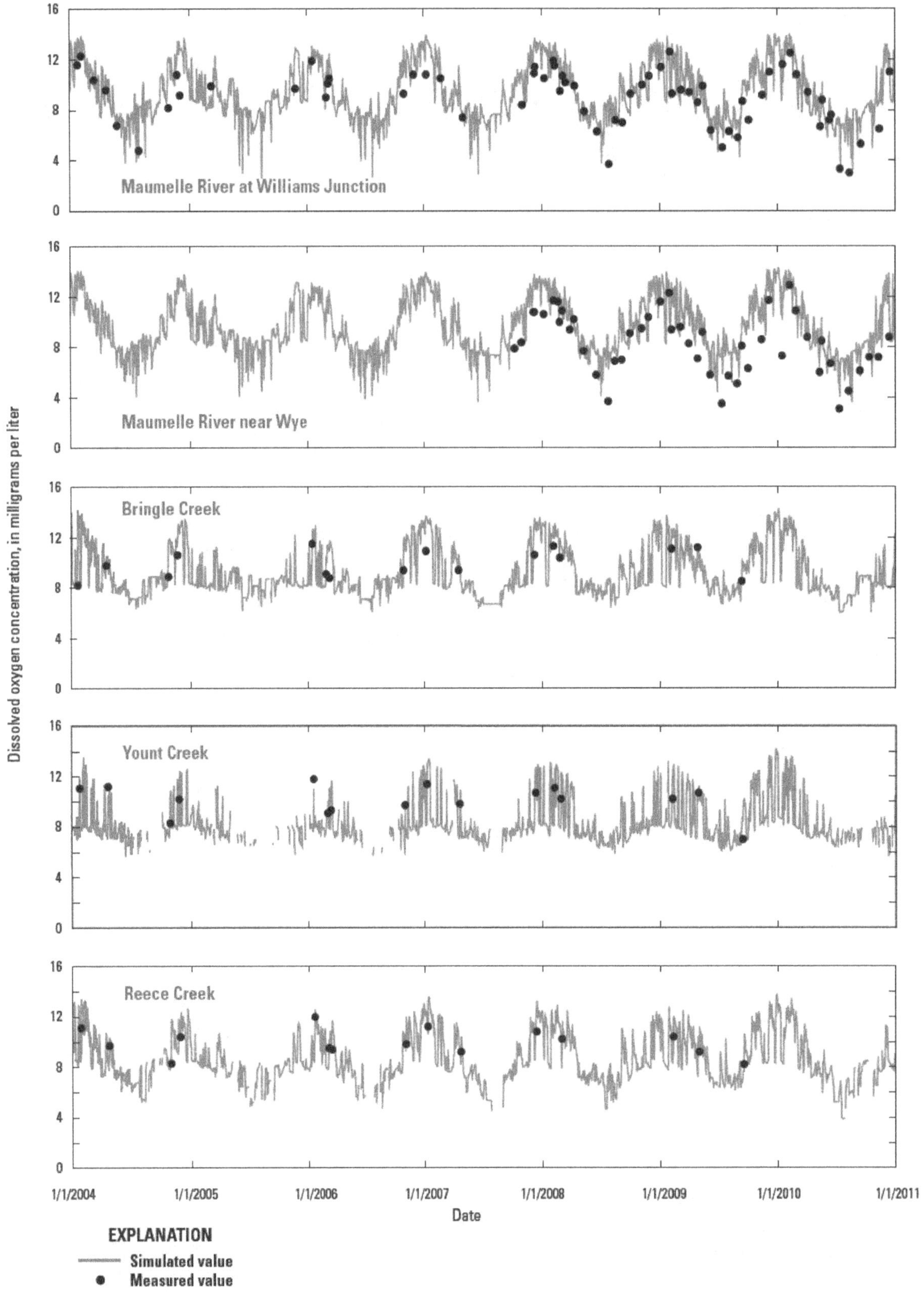

Figure 13.   Simulated and measured dissolved oxygen concentrations at selected inflow stations in the Lake Maumelle watershed.

The simulated values for fecal coliform bacteria during periods of base flow (streamflows not substantially influenced by runoff) agree reasonably well for Williams Junction (fig. 14) (with differences generally ranging from -1,188 to 21 col/100 mL, and percent difference generally ranging from -100 to 58 percent) and Wye (with differences generally ranging from -851 to 15 col/100 mL, and percent difference ranging from -100 to 13 percent). Additionally, simulated fecal coliform bacteria matched well with the quarterly and monthly sampling values and also during periods of stormflow (streamflow substantially influenced by runoff) for most stations (fig. 14). For paired (measured values and simulated daily mean values for days with measured values) fecal coliform bacteria concentrations at Williams Junction, the mean measured fecal coliform bacteria concentration was 389 col/100 mL and the mean daily simulated fecal coliform bacteria concentration was 105 col/100 mL (table 10). The RPDs between the annual S–LOADEST and simulated HSPF fecal coliform bacteria loads ranged from 59.90 to 122.39 percent with a median value of 79.30 percent (table 11).

The median (2004 through 2010) loading rates for fecal coliform bacteria for subwatersheds draining directly into Lake Maumelle (the rates are calculated from the area upstream from each of these subwatersheds plus the area in the subwatershed) ranged from 1.31×10^9 to 1.60×10^{10} (col/acre)/yr (table 12). The median fecal coliform bacteria loading rate for subwatershed 22 (the area upstream from the mouth of the Maumelle River; the largest subwatershed in the Lake Maumelle watershed) was 6.18×10^9 (col/acre)/yr.

Nutrient Calibration

The mechanism for movement of nutrients is the same mechanism for movement of water within HSPF. Nutrients can be transported to a reach from a PERLND through overland flow, interflow, or groundwater discharge and through overland flow from an IMPLND. Dissolved nitrite plus nitrate nitrogen and dissolved ammonia nitrogen were simulated with accumulation (ACQOP), storage (SQOLIM), and washoff (WSQOP) parameters. The WSQOP parameter is the rate of surface runoff that will remove 90 percent of the stored quality constituent on the PERLND and IMPLND land segments; the units are inches per hour. Dissolved nitrite plus nitrate nitrogen and dissolved ammonia nitrogen were considered an overland flow constituent with contribution from interflow and groundwater. Additionally, dissolved nitrite plus nitrate nitrogen and dissolved ammonia nitrogen concentrations were applied to the PERLNDs, IMPLNDs, and stream reaches through wet atmospheric deposition (precipitation). These data were obtained from the National Atmospheric Deposition Program station at Caddo Valley, Ark. (National Atmospheric Deposition Program, 2012) (fig. 5). Dissolved orthophosphate was modeled as a sediment-associated constituent, in that the transport of the nutrient from the land surface to the reach is by attachment and removal of sediment particles. Additionally, a washoff potency factor (POTFW) was applied during periods of surface runoff for dissolved orthophosphate. The HSPF model was calibrated against nutrient concentrations for all five inflow stations. The WSQOP parameter (table 9) had the greatest effect on the nutrient calibration of all the water-quality parameters.

Phosphorus Calibration

The movement of phosphorus from the land surface to the stream reach was simulated using the generalized quality constituent module (GQUAL) for PERLNDs and IMPLNDs. Dissolved orthophosphate movement from the land surface was modeled as a sediment-associated constituent with 10 percent, 30 percent, and 60 percent of the sediment being partitioned to sand, silt, and clay, respectively. A GQUAL that is sediment-associated considers the advection of adsorbed suspended material, deposition and scour of adsorbed material with sediment, decay of suspended sediment and bed material, and adsorption/desorption between the dissolved and sediment-associated phase. Both the dissolved and total forms of phosphorus were simulated by the HSPF watershed model. However, only the dissolved orthophosphate values simulated by the HSPF model were provided as input into the CE–QUAL–W2 model because the underlying equations require only the dissolved form of phosphorus.

The simulated values for dissolved orthophosphate concentrations during periods of base flow (streamflows not substantially influenced by runoff) agree reasonably well for Williams Junction (fig. 15) (with differences generally ranging from -0.006 to 8.92×10^{-5} mg/L, and percent difference generally ranging from -82 to 2.2 percent) and Wye (with differences generally ranging from -0.004 to 0.002 mg/L, and percent difference ranging from -53 to 54 percent). Additionally, simulated dissolved orthophosphate concentrations matched well with the quarterly and monthly sampling values and also during periods of stormflow (streamflow substantially influenced by runoff) for all stations (fig. 15). Several of the measured dissolved orthophosphate values for all inflow stations were less than the LRL of 0.006 mg/L. For paired (measured values and simulated daily mean values for days with measured values) dissolved orthophosphate concentrations at Williams Junction, the mean measured dissolved orthophosphate concentration was less than 0.008 mg/L, and the mean daily simulated dissolved orthophosphate concentration was 0.003 mg/L (table 10). The RPDs between the annual S–LOADEST and simulated HSPF dissolved orthophosphate loads ranged from 34.62 to 114.68 percent with a median value of 74.39 percent (table 11).

The median (2004 through 2010) loading rates for dissolved orthophosphate for subwatersheds draining directly into Lake Maumelle (the rates are calculated from the area upstream from each of these subwatersheds plus the area in the subwatershed) ranged from 0.003 to 0.037 (lb/acre)/yr as P (table 12). The median dissolved orthophosphate loading rate for subwatershed 22 (the area upstream from the mouth of the Maumelle River; the largest subwatershed in the Lake Maumelle watershed) was 0.037 (lb/acre)/yr as P.

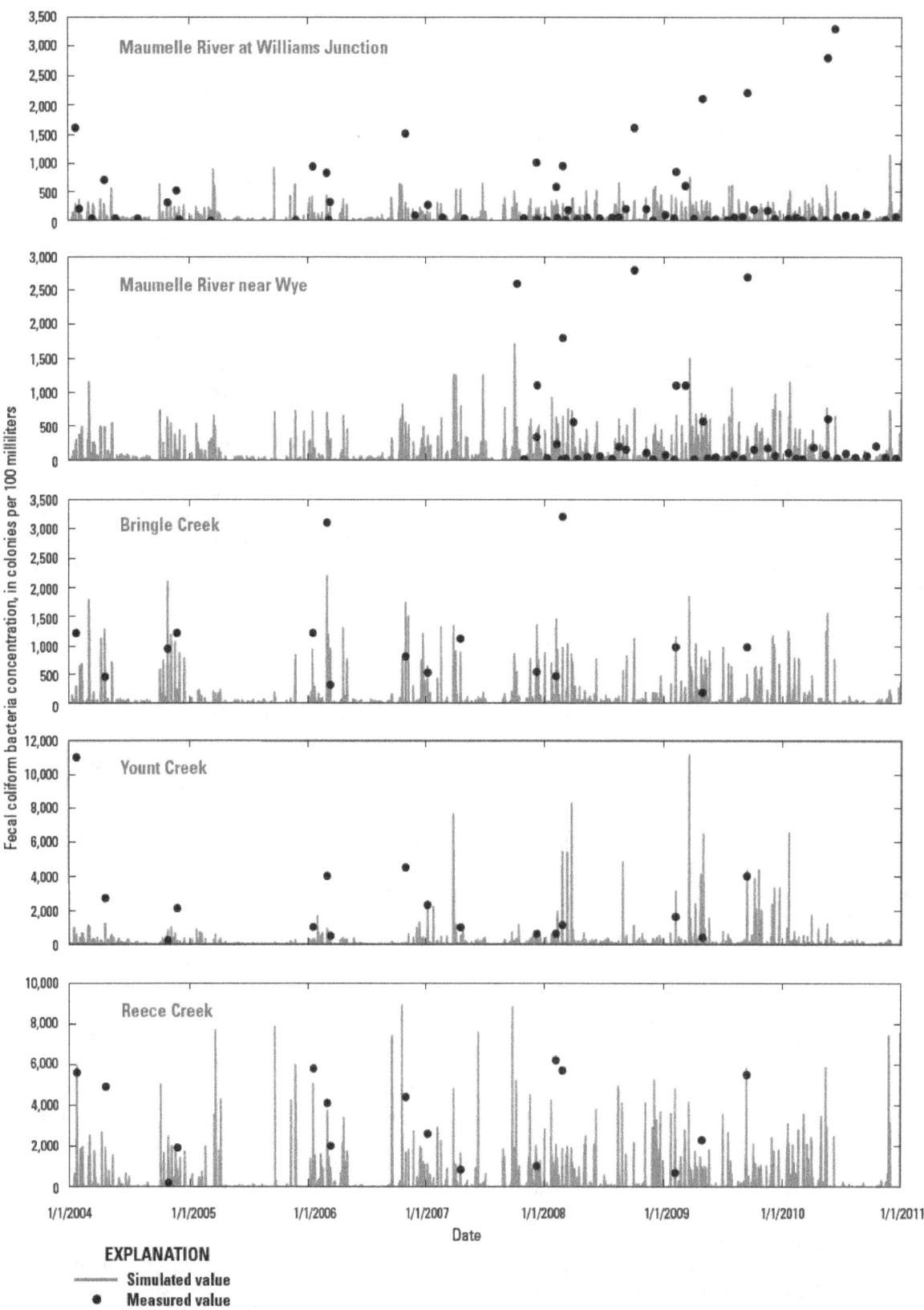

Figure 14.   Simulated and measured fecal coliform bacteria concentrations at selected inflow stations in the Lake Maumelle watershed.

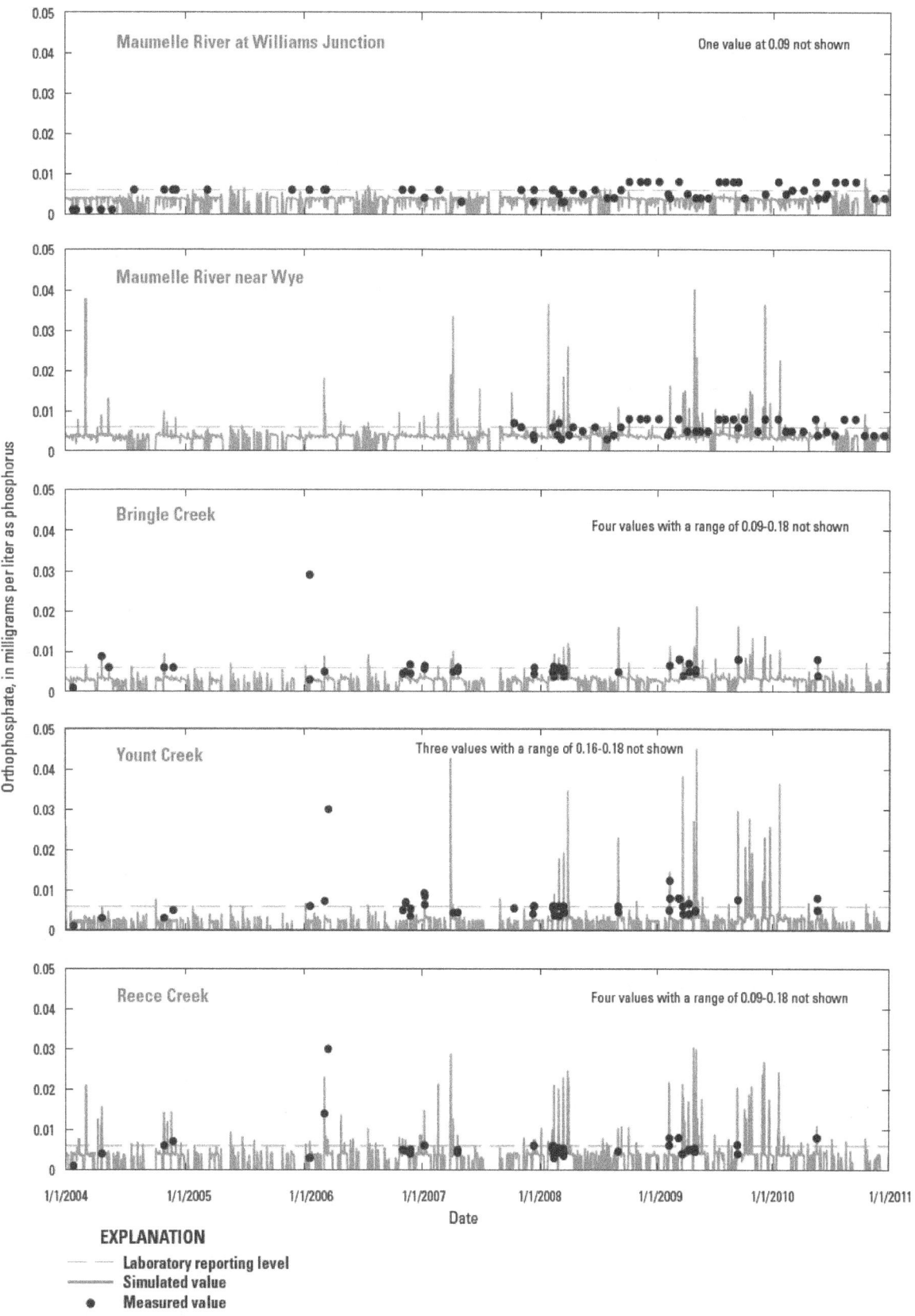

Figure 15. Simulated and measured dissolved orthophosphate concentrations at selected inflow stations in the Lake Maumelle watershed.

The simulated values for total phosphorus concentrations during periods of base flow (streamflows not substantially influenced by runoff) agree reasonably well for Williams Junction (fig. 16) (with differences generally ranging from -0.022 to 0.077 mg/L, and percent difference ranging from -53 to 200 percent) and Wye (with differences generally ranging from -0.054 to -0.006 mg/L, with percent difference ranging from -92 to -62 percent). Additionally, simulated total phosphorus concentrations matched well with the quarterly and monthly sampling values and also during periods of stormflow (streamflow substantially influenced by runoff) for all stations (fig. 16). For paired (measured values and simulated daily mean values for days with measured values) total phosphorus concentrations at Williams Junction, the mean measured total phosphorus concentration was 0.032 mg/L, and the mean daily simulated total phosphorus concentration was 0.043 mg/L (table 10). The RPDs between the annual S–LOADEST and simulated HSPF total phosphorus loads ranged from 22.12 to 85.96 percent with a median value of 48.45 percent (table 11).

The median (2004 through 2010) loading rates for total phosphorus for subwatersheds draining directly into Lake Maumelle (the rates are calculated from the area upstream from each of these subwatersheds plus the area in the subwatershed) ranged from 0.139 to 0.355 (lb/acre)/yr (table 12). The median total phosphorus loading rate for subwatershed 22 (the area upstream from the mouth of the Maumelle River; the largest subwatershed in the Lake Maumelle watershed) was 0.355 (lb/acre)/yr.

Dissolved Nitrite plus Nitrate Nitrogen Calibration

The movement of dissolved nitrite plus nitrate nitrogen from the land surface to the reach was simulated through the GQUAL module for PERLNDs and IMPLNDs; therefore, only advection and a generalized first-order decay process were considered. The greatest contribution of nitrate plus nitrite to the system comes from wet atmospheric deposition. This was determined through sensitivity analysis of the HSPF model and also through the results of the USGS SPAtially-Referenced Regression on Watershed attributes (SPARROW) model. When nitrate plus nitrite wet atmospheric deposition was removed from the HSPF model, little to no nitrogen was transported through the system. U.S. Geological Survey (2012) estimated that 78.8 percent of the nitrogen load in the Lake Maumelle watershed comes from atmospheric deposition.

The simulated values for dissolved nitrite plus nitrate nitrogen concentrations during periods of base flow (streamflows not substantially influenced by runoff) agree reasonably well for Williams Junction (fig. 17) (with differences generally ranging from -0.096 to 0.013 mg/L, with percent difference generally ranging from -83 to 95 percent) and Wye (with differences generally ranging from -0.063 to 0.008 mg/L, with percent difference generally ranging from -77 to 75 percent). Additionally, simulated dissolved nitrite plus nitrate nitrogen concentrations matched reasonably well with the quarterly and monthly sampling values and also during periods of stormflow (streamflow substantially influenced by runoff) for most stations (fig. 17). For paired (measured values and simulated daily mean values for days with measured values) dissolved nitrite plus nitrate nitrogen concentrations at Williams Junction, the mean measured dissolved nitrite plus nitrate nitrogen concentration was 0.059 mg/L, and the mean daily simulated dissolved nitrite plus nitrate nitrogen concentration was 0.030 mg/L (table 10). The RPDs between the annual S–LOADEST and simulated HSPF dissolved nitrite plus nitrate nitrogen loads ranged from 0.01 to 95.37 percent with a median value of 39.75 percent (table 11).

The median (2004 through 2010) loading rates for dissolved nitrite plus nitrate for subwatersheds draining directly into Lake Maumelle (the rates are calculated from the area upstream from each of these subwatersheds plus the area in the subwatershed) ranged from 0.081 to 0.182 (lb/acre)/yr as N (table 12). The median dissolved nitrite plus nitrate loading rate for subwatershed 22 (the area upstream from the mouth of the Maumelle River; the largest subwatershed in the Lake Maumelle watershed) was 0.178 (lb/acre)/yr as N.

Dissolved Ammonia Nitrogen Calibration

The movement of dissolved ammonia nitrogen from the land surface to the reach was simulated through the GQUAL module for PERLNDs and IMPLNDs; therefore, only advection and a generalized first-order decay process were considered. The greatest contribution of dissolved ammonia nitrogen to the system comes from wet atmospheric deposition. This also was determined through sensitivity analysis and the results of the SPARROW model (see discussion in "Dissolved Nitrite plus Nitrate Nitrogen Calibration") (U.S. Geological Survey, 2012).

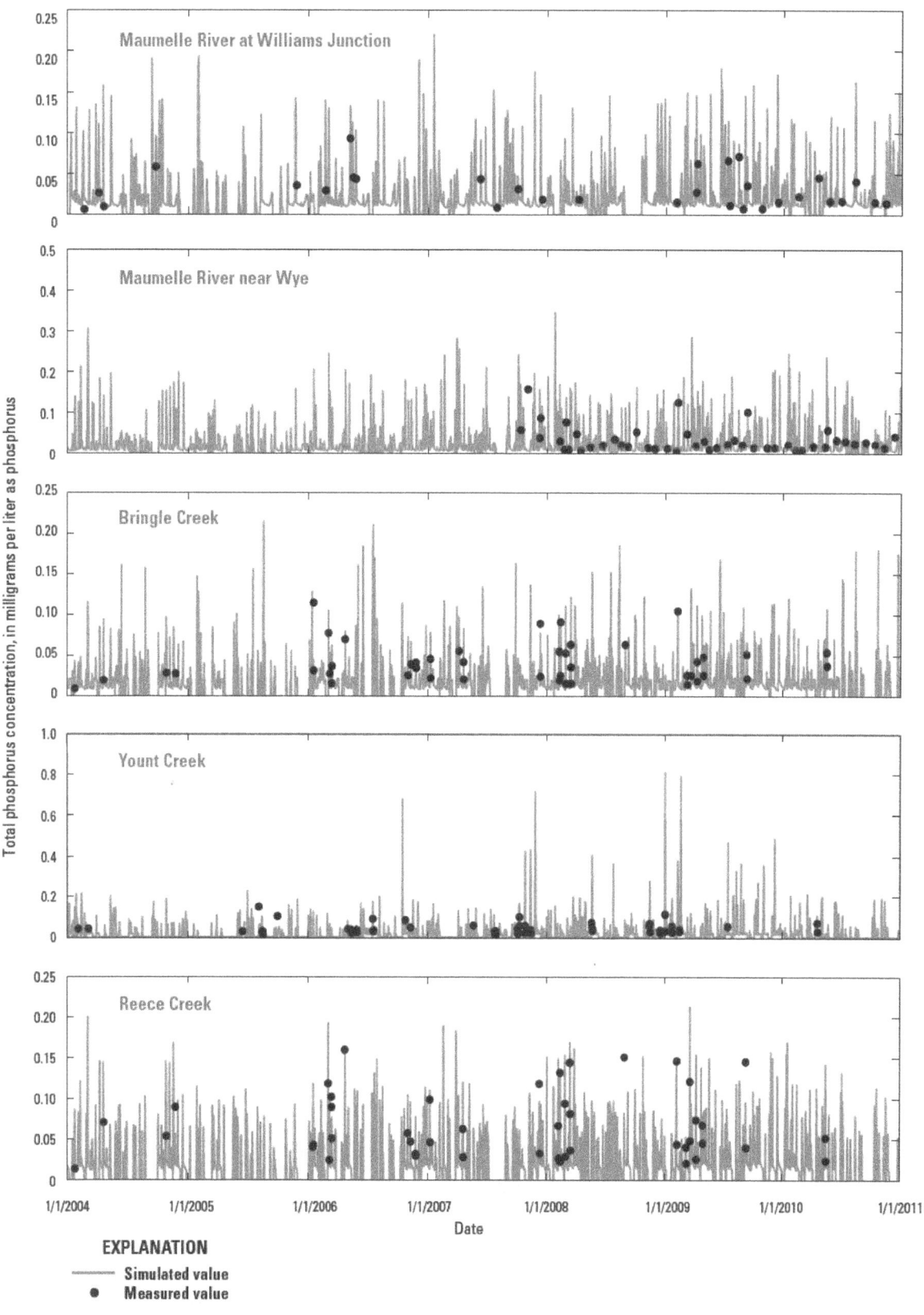

Figure 16. Simulated and measured total phosphorus concentrations at selected inflow stations in the Lake Maumelle watershed.

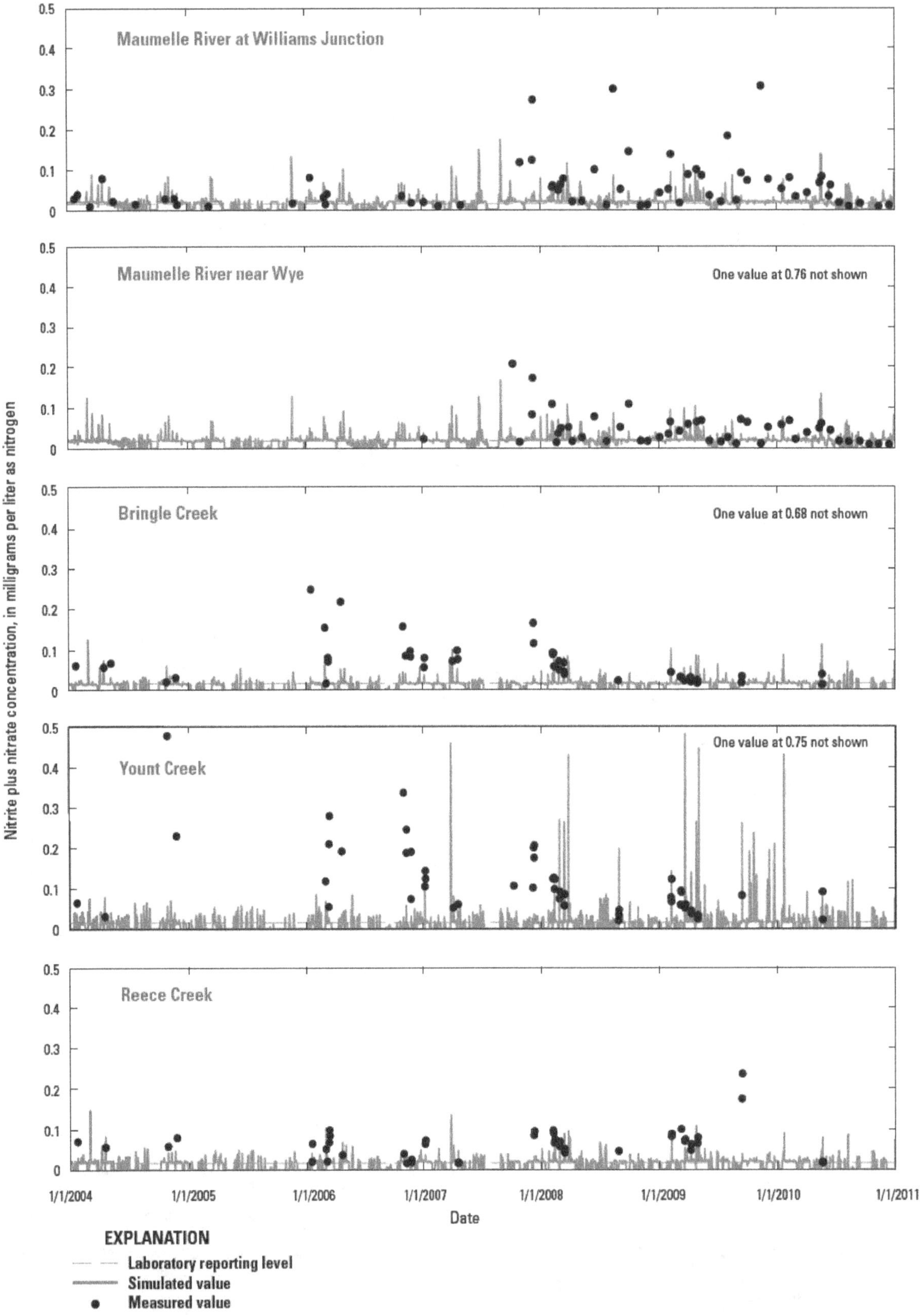

Figure 17.   Simulated and measured nitrite plus nitrate concentrations at selected inflow stations in the Lake Maumelle watershed.

The simulated values for dissolved ammonia nitrogen concentrations during periods of base flow (streamflows not substantially influenced by runoff) agree reasonably well for Williams Junction (fig. 18) (with differences generally ranging from -0.020 to 0.033 mg/L, and percent difference generally ranging from -100 to 406 percent) and Wye (with differences generally ranging from -0.019 to 0.010 mg/L, and percent difference generally ranging from -100 to 89 percent). Additionally, simulated dissolved ammonia nitrogen concentrations matched well with the quarterly and monthly sampling values and also during periods of stormflow (streamflow substantially influenced by runoff) for most stations (fig. 18). For paired (measured values and simulated daily mean values for days with measured values) dissolved ammonia nitrogen concentrations at Williams Junction, the mean measured dissolved ammonia nitrogen concentration was less than 0.02 mg/L, and the mean daily simulated dissolved ammonia nitrogen concentration was 0.017 mg/L (table 10). The RPDs between the annual S–LOADEST and simulated HSPF dissolved ammonia nitrogen loads ranged from 16.56 to 104.43 percent with a median value of 91.19 percent (table 11).

The median (2004 through 2010) loading rates for dissolved ammonia for subwatersheds draining directly into Lake Maumelle (the rates are calculated from the area upstream from each of these subwatersheds plus the area in the subwatershed) ranged from 0.039 to 0.138 (lb/acre)/yr as N (table 12). The median dissolved ammonia loading rate for subwatershed 22 (the area upstream from the mouth of the Maumelle River; the largest subwatershed in the Lake Maumelle watershed) was 0.130 (lb/acre)/yr as N.

Total Organic Carbon Calibration

Total organic carbon was modeled in a manner similar to fecal coliform bacteria; it was modeled as a general quality constituent where there is accumulation and storage on the land surface. Total organic carbon enters a stream reach by direct runoff. Therefore, in HSPF, total organic carbon is accumulated on the land surface during dry periods and washed off into stream reaches during wet periods. Washoff rates are adjusted to control the removal of 90 percent of the constituent (total organic carbon) per hour on the surface into the stream reach (WSQOP). Once in the stream reach, total organic carbon has a first-order decay rate applied to simulate degradation, assimilation, and other losses.

The simulated values for total organic carbon concentration during periods of base flow (streamflows not substantially influenced by runoff) agree reasonably well for Williams Junction (fig. 19) (with differences generally ranging from -7 to 7 mg/L, and percent difference generally ranging from -84 to 112 percent) and Wye (with differences generally ranging from -5 to 5 mg/L, and percent difference generally ranging from -91 to 76 percent). Additionally, simulated total organic carbon concentration matched well with the quarterly and monthly sampling values and also during periods of stormflow (streamflow substantially influenced by runoff) for all stations (fig. 19). For paired (measured values and simulated daily mean values for days with measured values) total organic carbon concentrations at Williams Junction, the mean measured total organic carbon concentration was 5.3 mg/L, and the mean daily simulated total organic carbon concentration was 4.1 mg/L (table 10). The RPDs between the annual S–LOADEST and simulated HSPF total organic carbon loads ranged from 8.20 to 83.04 percent with a median value of 26.35 percent (table 11).

The median (2004 through 2010) loading rates for total organic carbon for subwatersheds draining directly into Lake Maumelle (the rates are calculated from the area upstream from each of these subwatersheds plus the area in the subwatershed) ranged from 11.7 to 30.5 (lb/acre)/yr (table 12). The median total organic carbon loading rate for subwatershed 22 (the area upstream from the mouth of the Maumelle River; the largest subwatershed in the Lake Maumelle watershed) was 30.5 (lb/acre)/yr.

CE–QUAL–W2 Reservoir Modeling Calibration

Successful reservoir model application requires model calibration that compares simulated results with measured conditions. The CE–QUAL–W2 model calibration was completed by adjusting parameters (table 4), within reasonable ranges, for the 7-year period from January 2004 to December 2010. As for the HSPF model, 2003 was used as a model warmup period. Calibration generally was achieved by first calibrating the water balance and water temperature, and then calibrating the water-quality conditions (dissolved oxygen, nutrients, and algae). Measured data from three lake stations (East of Highway 10, Little Italy, and Natural Steps; fig. 1) spatially and hydrologically representative of Lake Maumelle were used in the calibration process.

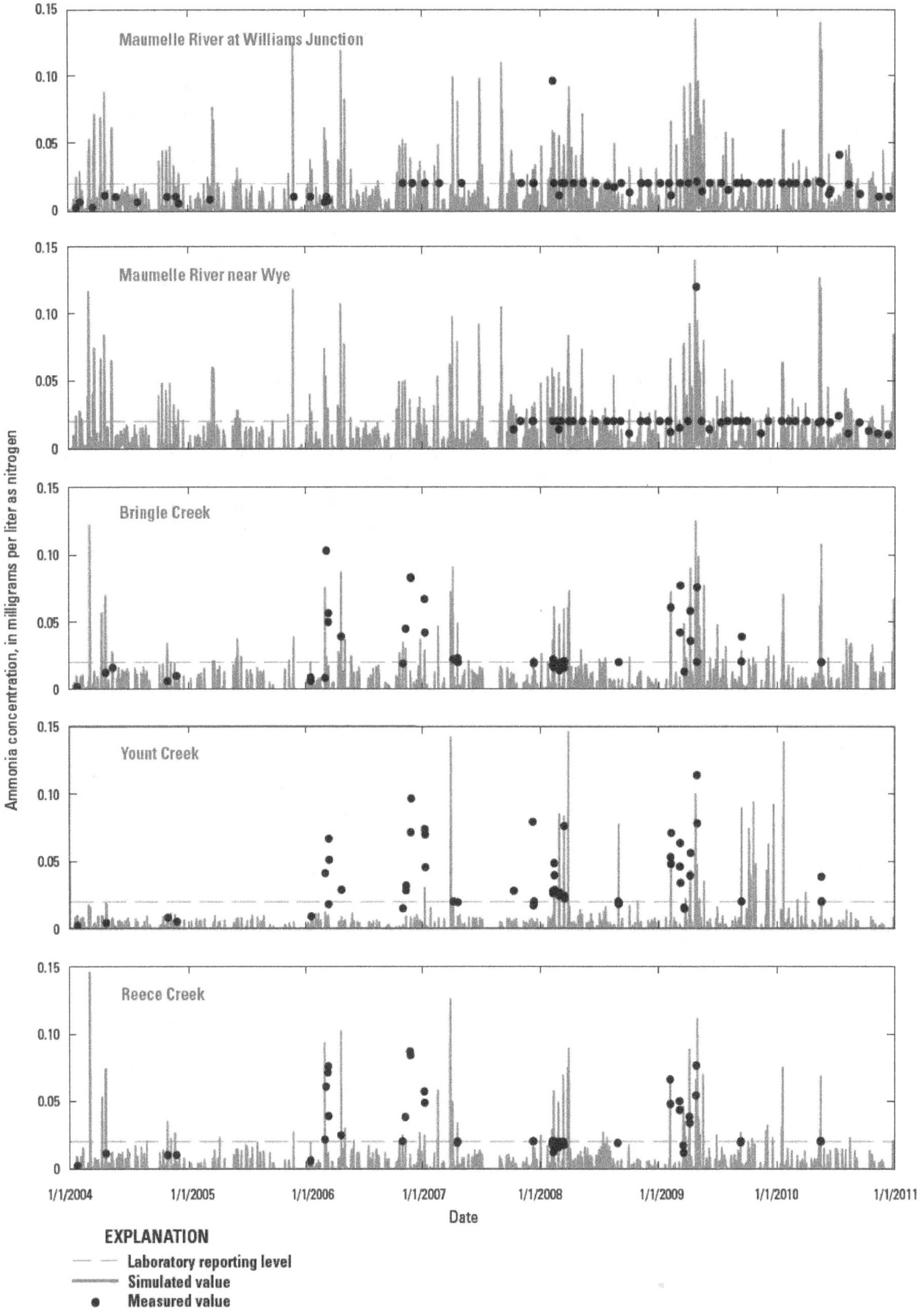

Figure 18.   Simulated and measured dissolved ammonia nitrogen concentrations at selected inflow stations in the Lake Maumelle watershed.

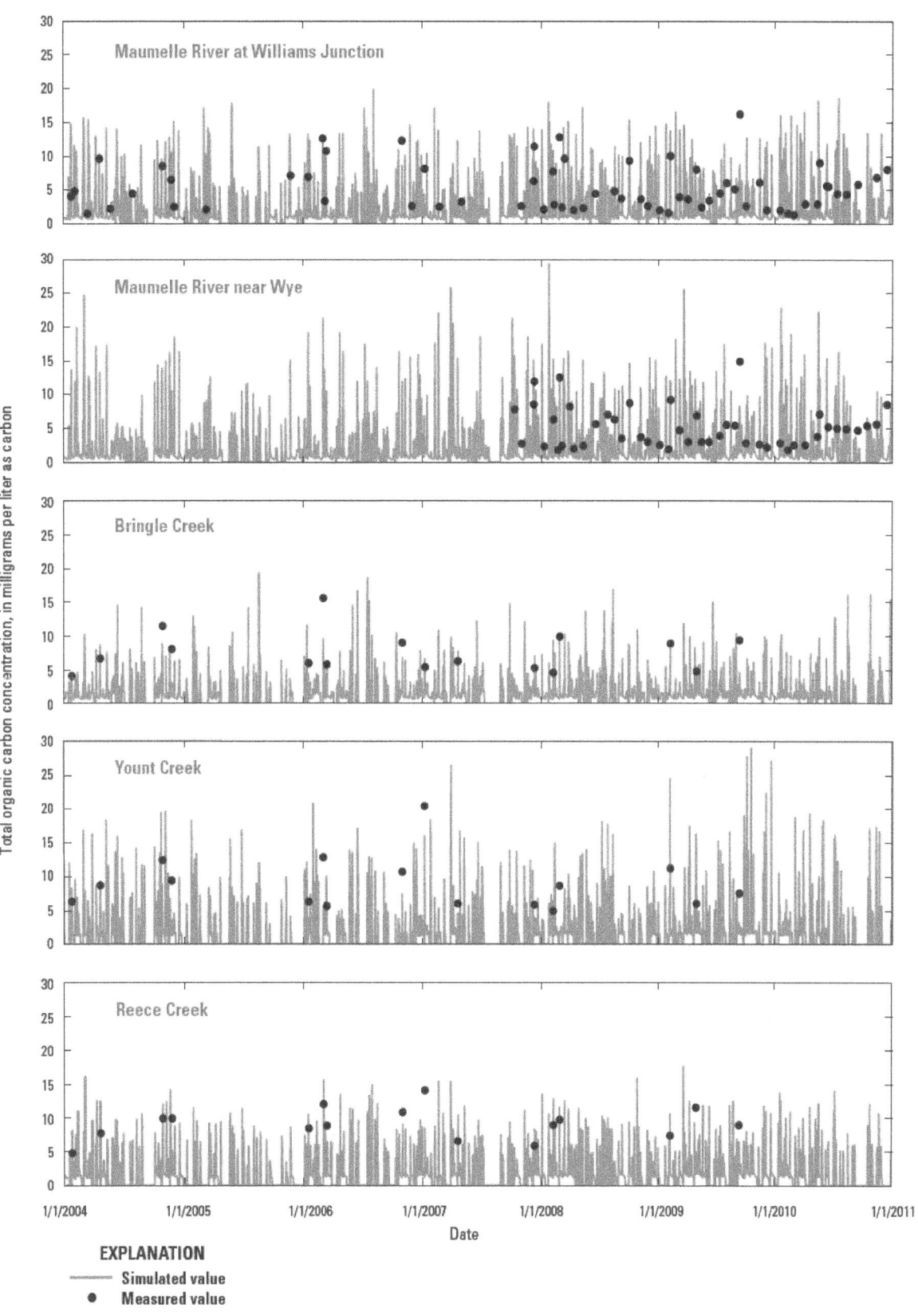

Figure 19.   Simulated and measured total organic carbon concentrations at selected inflow stations in the Lake Maumelle watershed.

Water Balance

Simulating the water-surface altitude associated with Lake Maumelle is important to accurately depict the hydrodynamics as the lake's volume changes in response to lake inflows. The HSPF model produced the simulated flow for all tributaries entering the lake and flow associated with overland flow for land surrounding the lake. The simulated inflows, precipitation on the pool, simulated outflow over the spillway, and measured drinking-water withdrawal provided the hydraulic boundary for the CE–QUAL–W2 model. Simulated water-surface altitudes in Lake Maumelle were adjusted to the measured water surface for the model period of January 1, 2004, and December 31, 2010 (fig. 20). The water-surface altitudes were corrected to the measured values by adjusting the ungaged values through the "distributed tributary" inflow file. Inflow was added or subtracted so that the simulated water-surface altitude reflected the measured water-surface altitude, therefore, accounting for unmeasured inflow and groundwater interaction in Lake Maumelle. Given the accurate water balance, temperature and water quality could be calibrated without the uncertainty incurred with having differences between simulated and measured lake volumes. The same "distributed tributary" inflow quantities for the calibrated baseline model were used for all three scenarios.

Water Temperature

Simulated water temperatures in the CE–QUAL–W2 model were compared to more than 60 depth profiles of temperature measured at each of the three stations on Lake Maumelle (East of Highway 10, Little Italy, and Natural Steps). During periods of thermal stratification, the epilimnion (the warm upper layer) and hypolimnion (the cool lower layer) are separated by a thermocline (layer of rapidly cooling water temperature with a small increase in depth). The terms "epilimnion" and "hypolimnion" refer to layers within a thermally stratified lake or a lake separated into thermal layers. A strong thermocline existed in Lake Maumelle at the Little Italy and Natural Steps stations approximately from May through September of all modeled years (fig. 21); a substantially weaker thermocline existed (for shorter periods of time) in the upstream part of Lake Maumelle at the East of Highway 10 station (fig. 21). Simulated vertical distributions of temperatures agreed with measured distributions even for complex temperature profiles. Although the calibrated model generally provided an excellent simulation of water temperature in Lake Maumelle, the simulation accuracy of water temperatures varied with water temperature, season, and depth. Occasionally, the depth of the simulated thermocline was a little above or below the measured thermocline and provided large differences between simulated and measured temperature within the layer of greatest change in temperature.

The largest differences between measured and simulated water temperature data occurred in the upstream part of the reservoir (segment 3, fig. 21), which is the most dynamic part of the reservoir. The upstream part of the reservoir is the shallowest section of Lake Maumelle and has more riverine characteristics than the deep lacustrine-type (lake-like) characteristics of the downstream part of the reservoir. The upstream part also receives most of the inflow to the reservoir, which creates more dynamic conditions. Much of the extreme error between individual measured and simulated water temperatures occurred in the region of depth where the thermocline existed, when present. Often the simulated thermocline depth was offset (higher or lower in the water column within a couple of inches) from where it was actually measured, causing large temperature differences at given depths. In most cases, if the simulated thermocline were positioned at the same depth as the measured thermocline, water temperatures would be similar, with little difference between two.

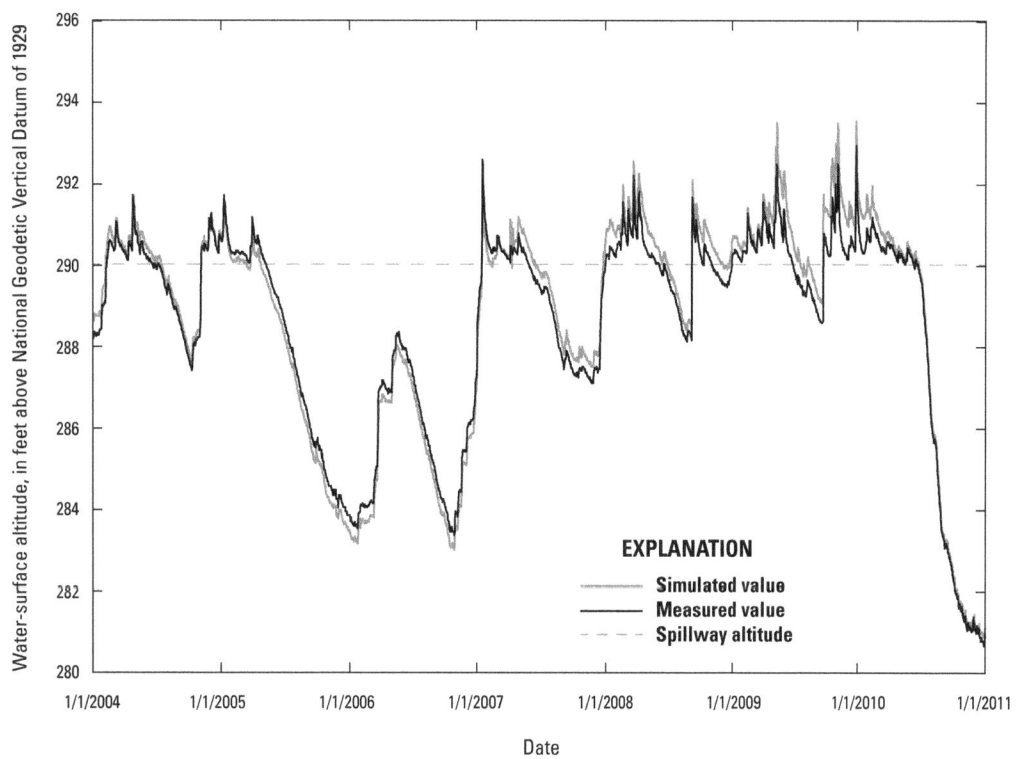

Figure 20.   Simulated and measured Lake Maumelle pool water-surface altitude.

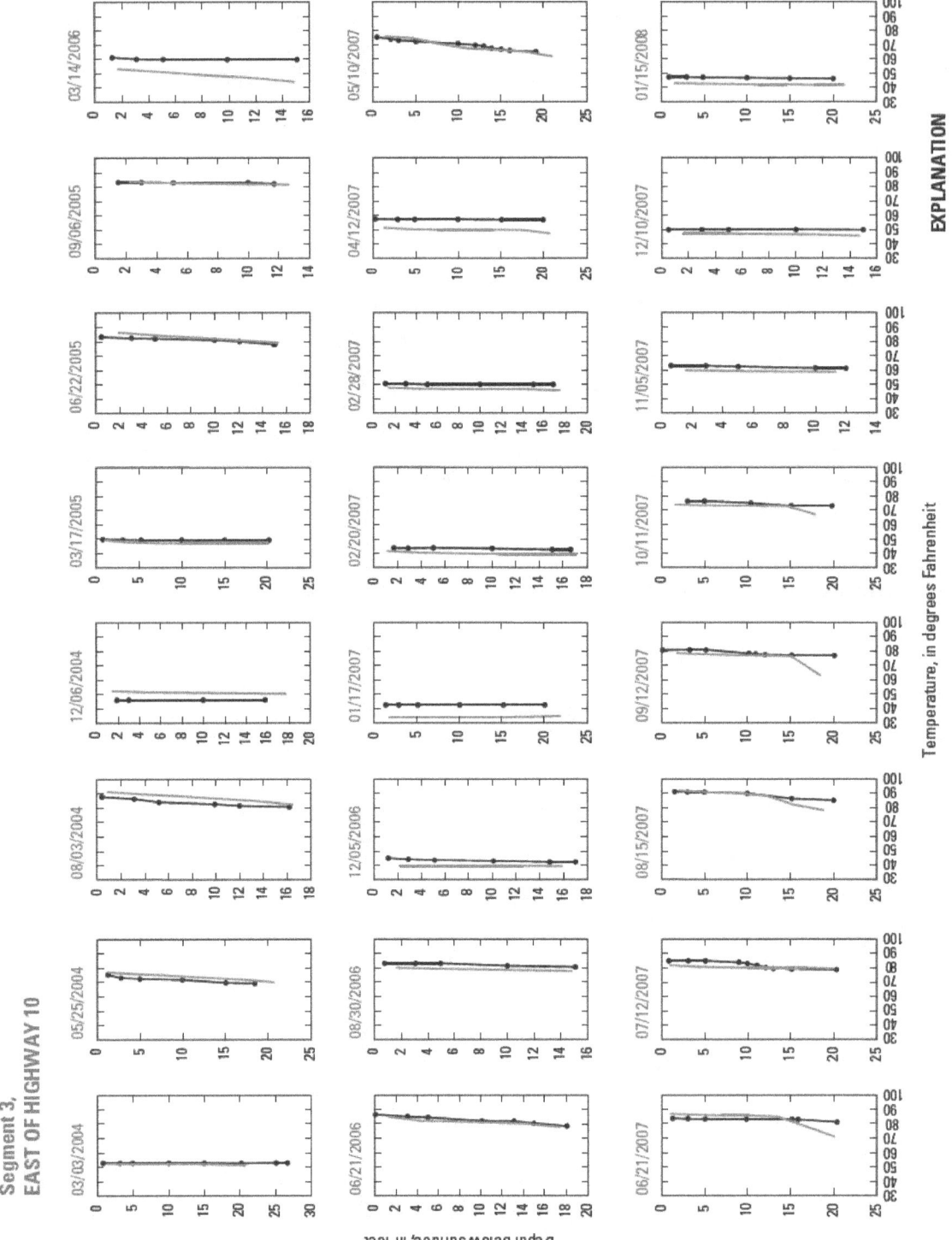

Figure 21.   Simulated and measured temperature-depth profiles for Lake Maumelle.

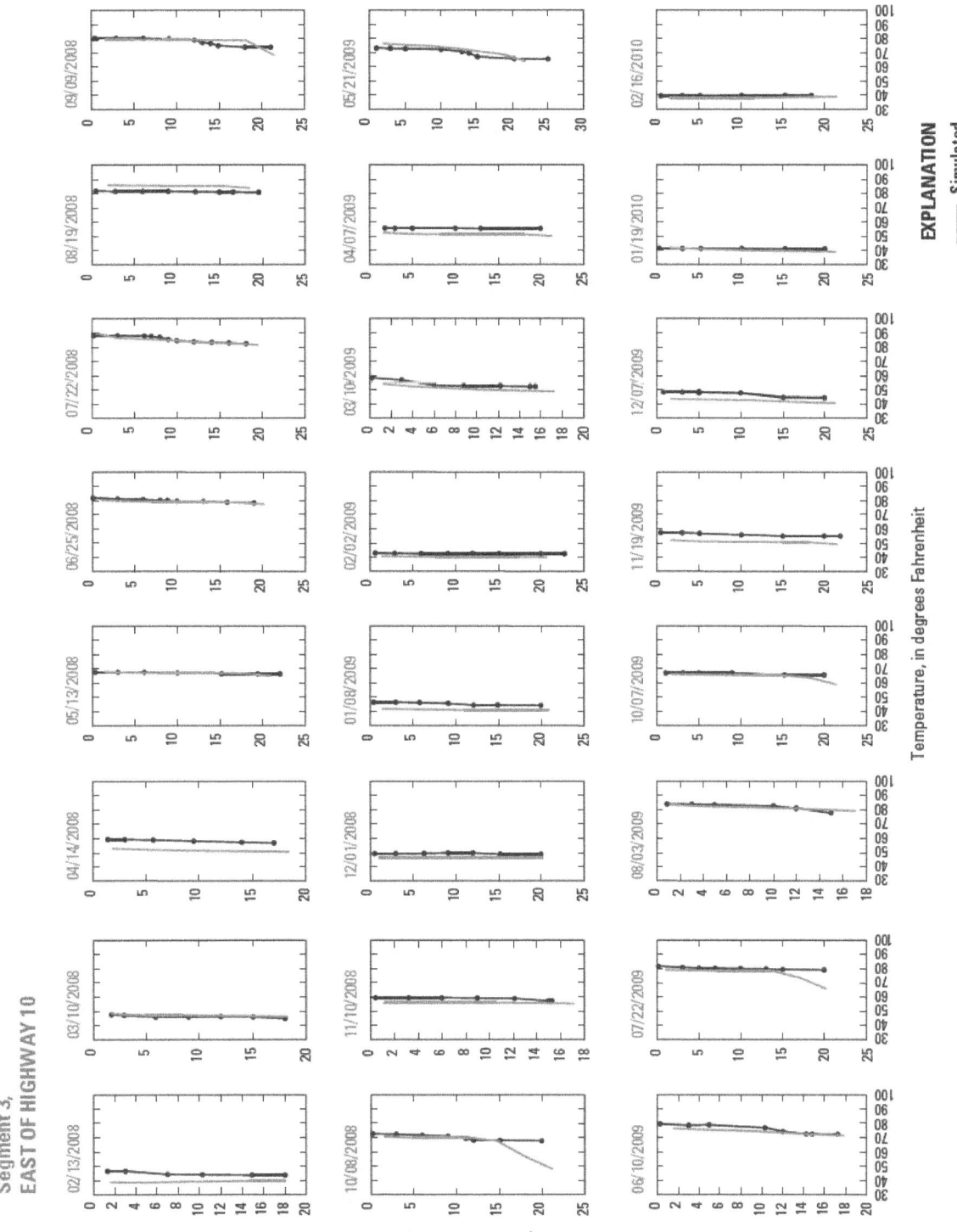

Figure 21. Simulated and measured temperature-depth profiles for Lake Maumelle.—Continued

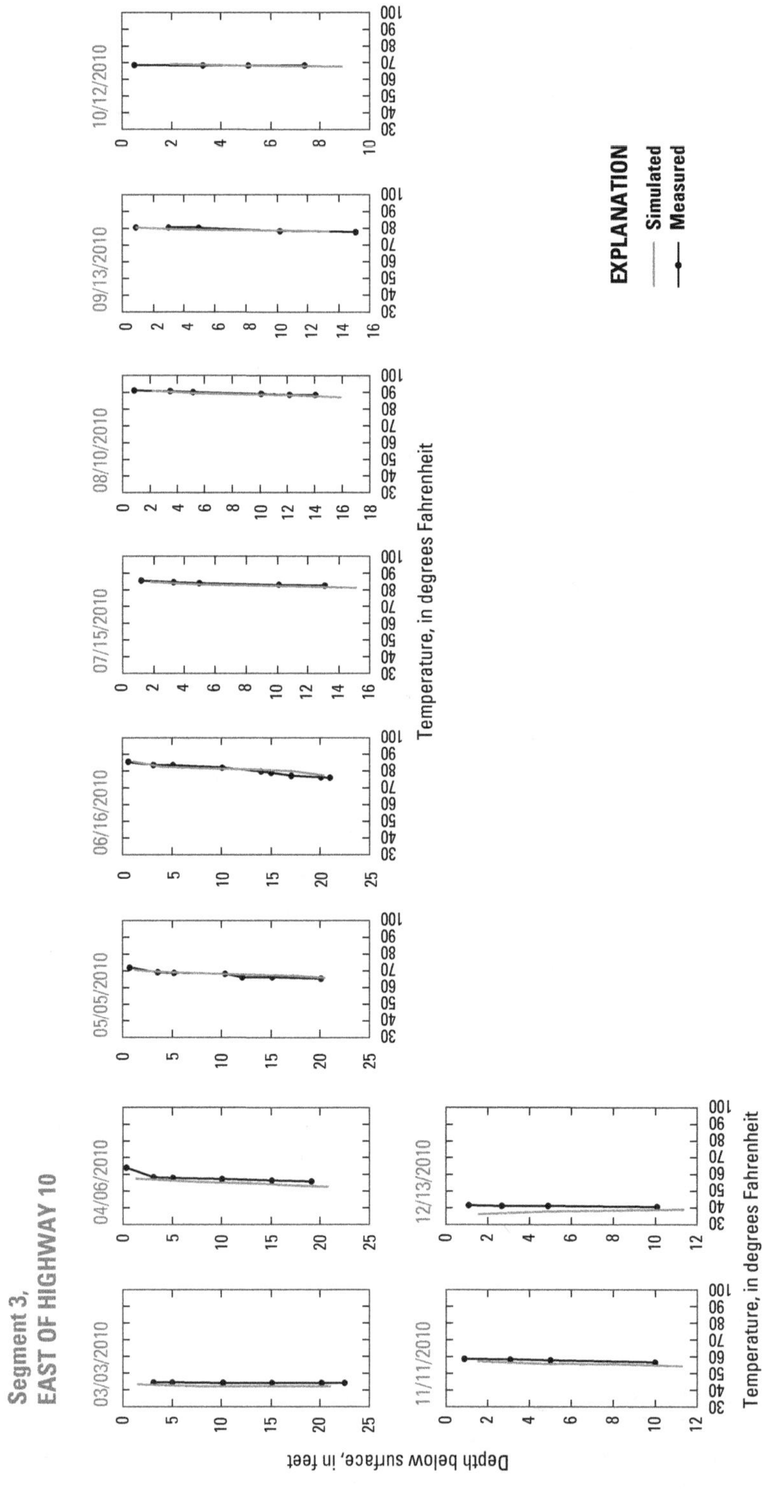

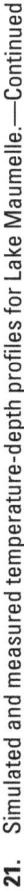

Figure 21. Simulated and measured temperature-depth profiles for Lake Maumelle.—Continued

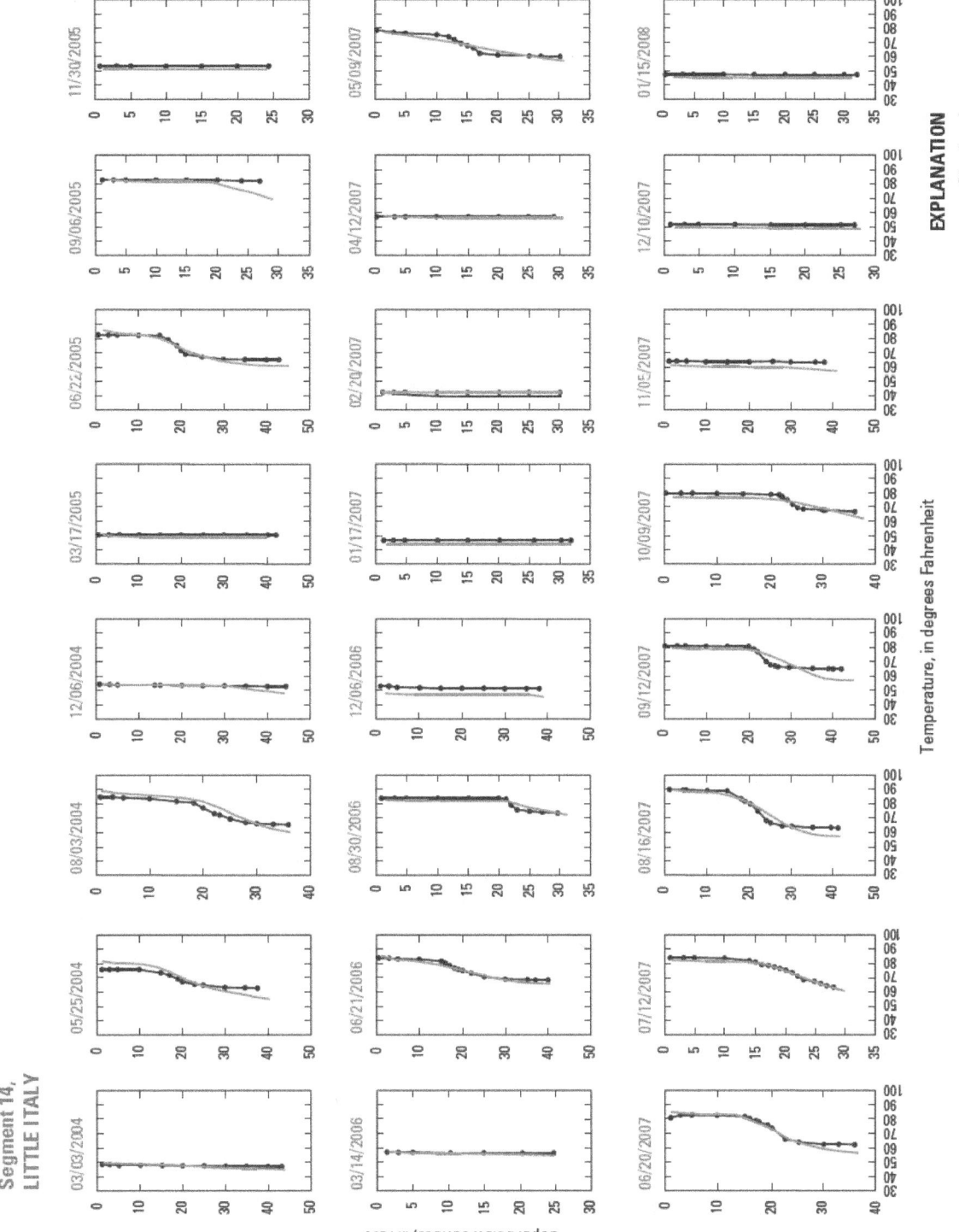

Figure 21. Simulated and measured temperature-depth profiles for Lake Maumelle.—Continued

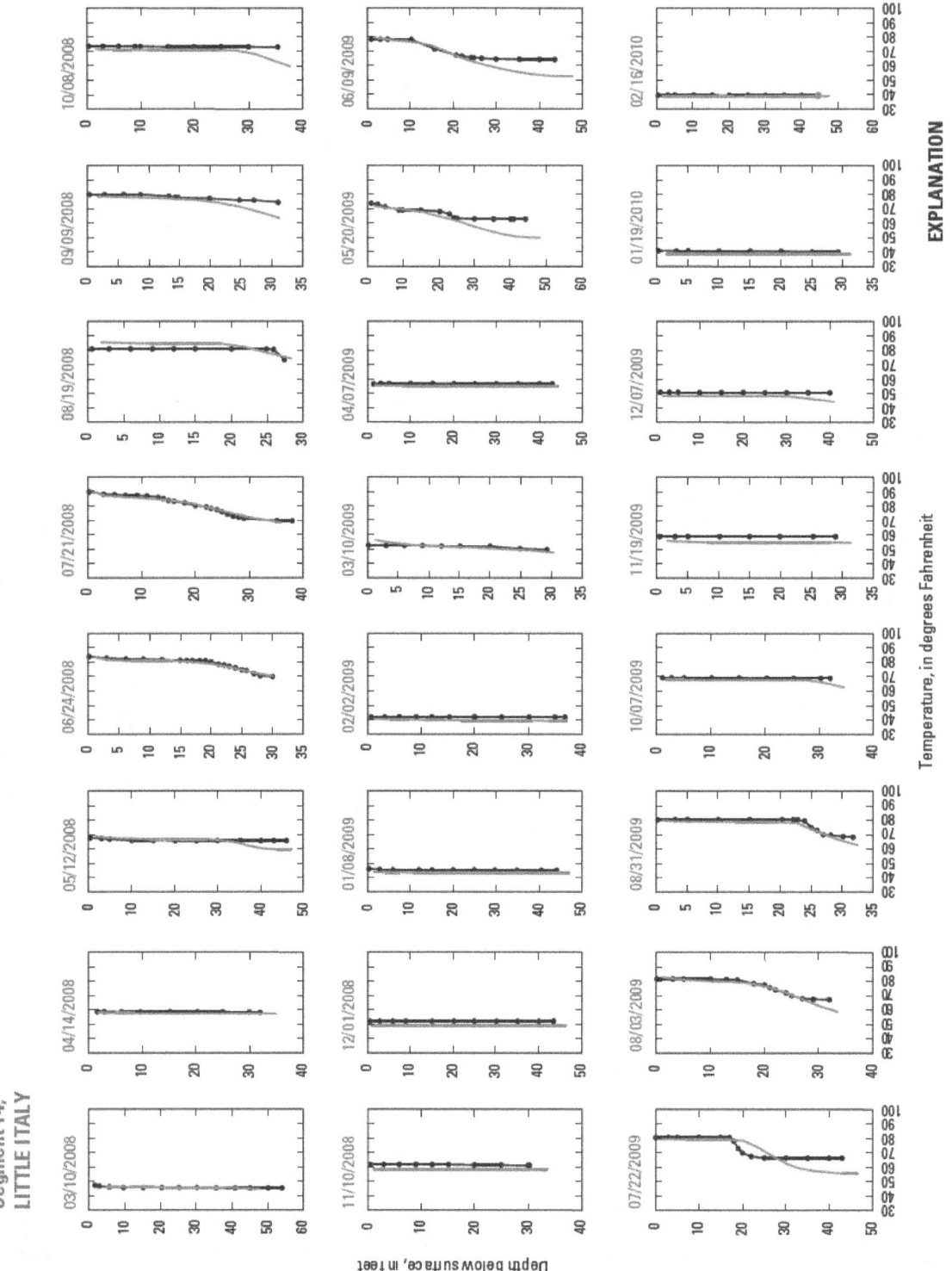

Figure 21. Simulated and measured temperature-depth profiles for Lake Maumelle.—Continued

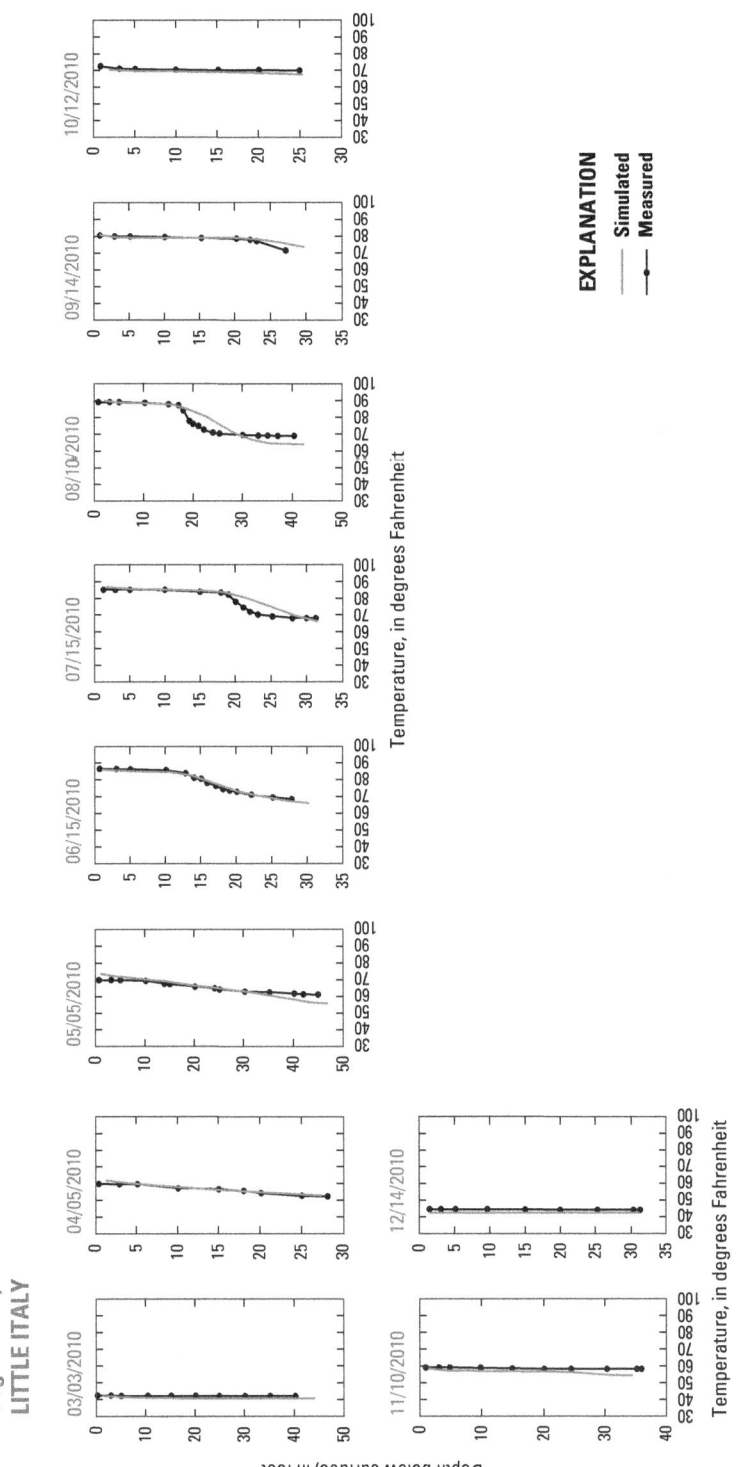

Figure 21. Simulated and measured temperature-depth profiles for Lake Maumelle.—Continued

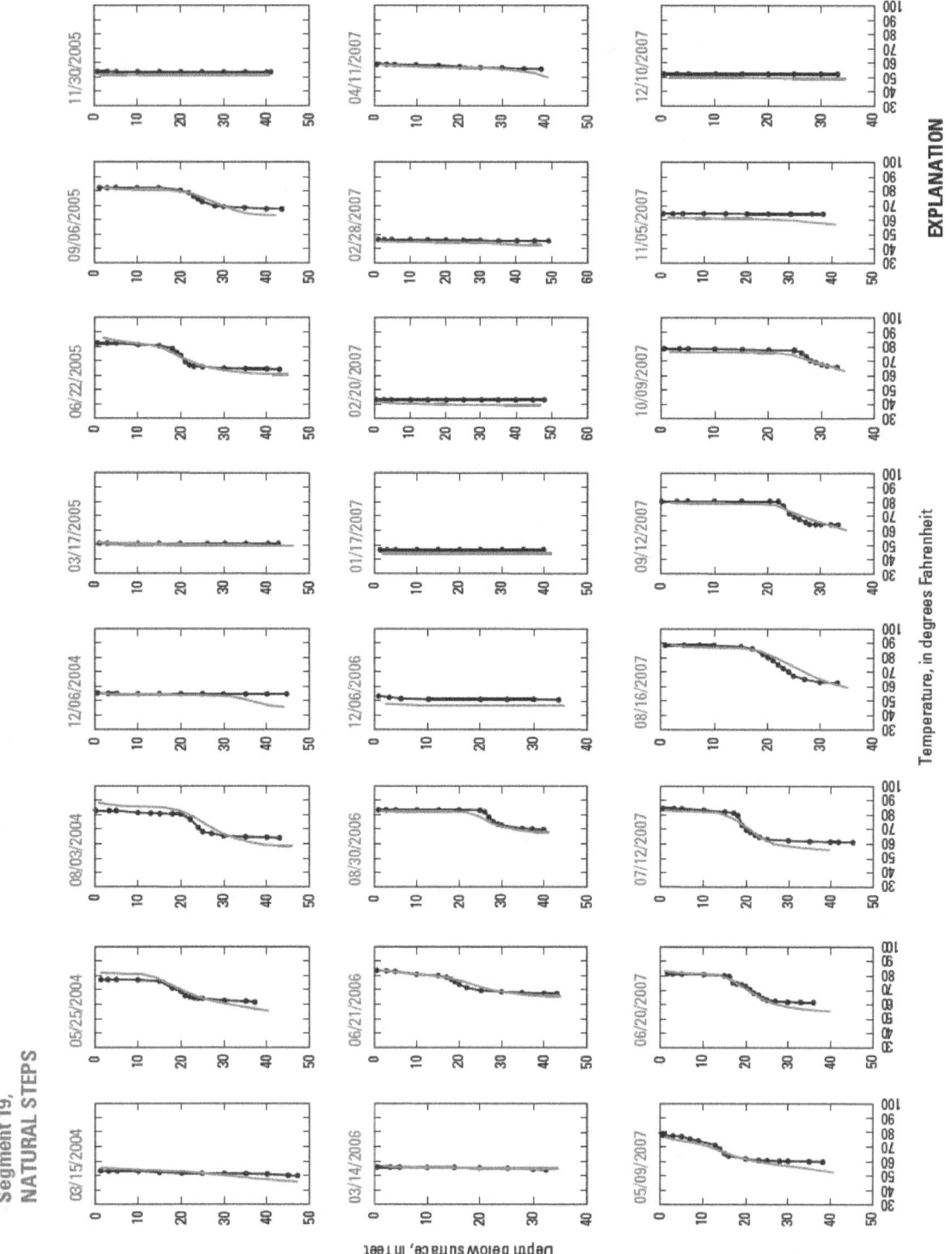

Figure 21.   Simulated and measured temperature-depth profiles for Lake Maumelle.—Continued

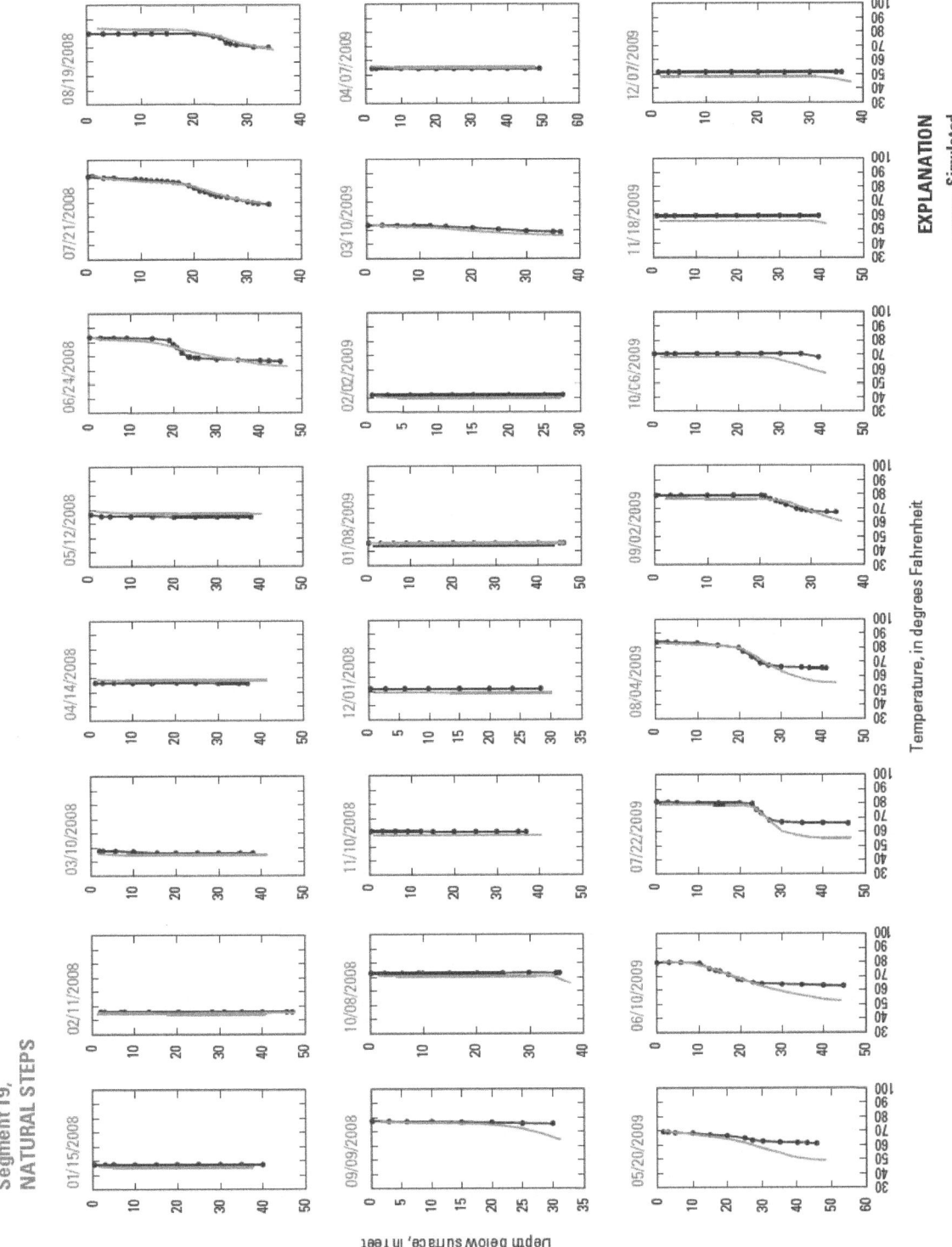

Figure 21. Simulated and measured temperature-depth profiles for Lake Maumelle.—Continued

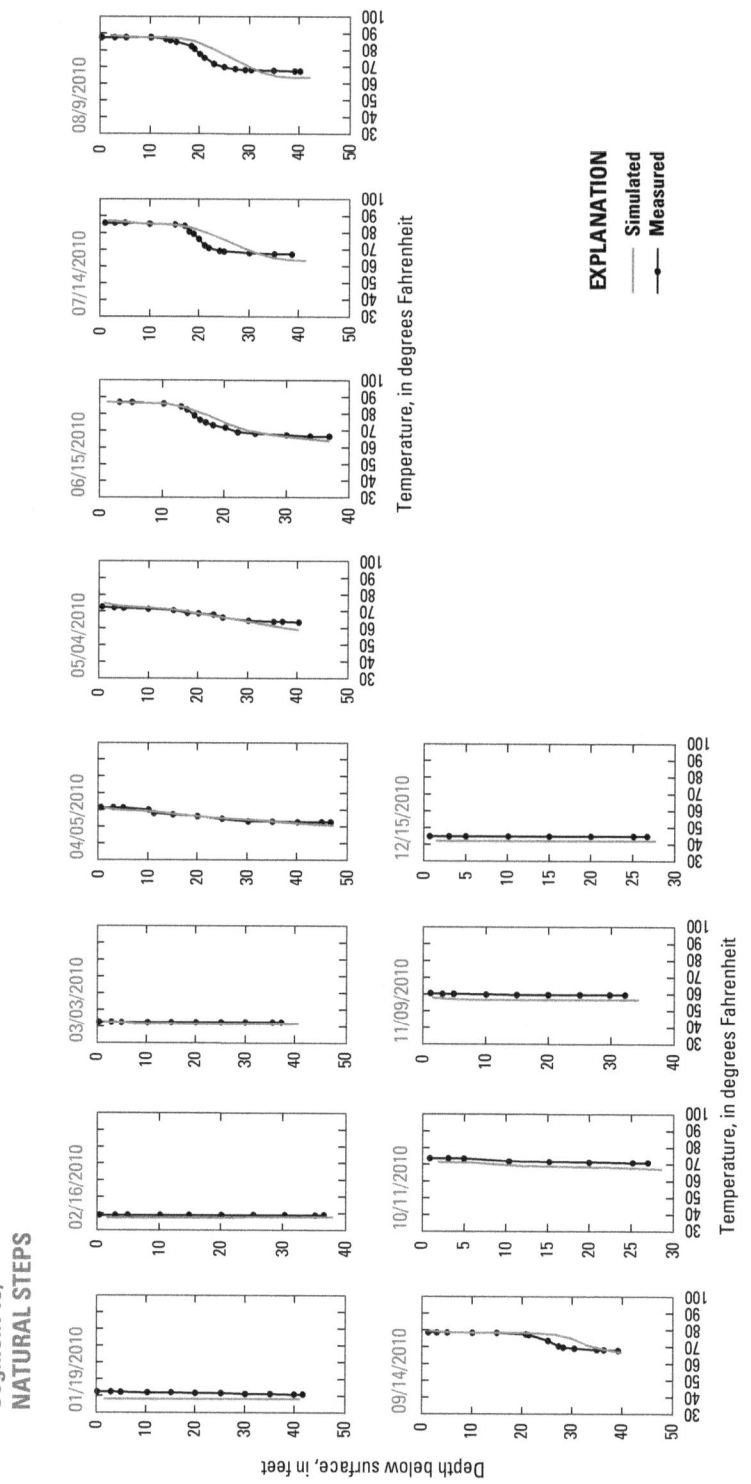

Figure 21.   Simulated and measured temperature-depth profiles for Lake Maumelle.—Continued

Results from the water temperature simulations provide information on physical characteristics and processes in the reservoir—information that might not be obtained from monthly or semimonthly measurements. Simulated water temperature ranged from 32.0 to 91.9°F for all stations (fig. 21). Mean differences between measured and simulated water temperature ranged from -1.07 to -0.49°F with a RMSE that ranged from 1.68 to 2.09°F and a MAE that ranged from 1.26 to 1.53°F for all stations over the simulation period (2004–10) (table 13).

Water Quality

Simulated dissolved oxygen concentrations were compared to measured depth profile data at three stations in Lake Maumelle. Other simulated water-quality constituent data were compared to samples collected in the epilimnion (3.28 ft (1 m) below the surface) and hypolimnion (3.28 ft (1 m) above the bottom).

Dissolved Oxygen

Simulated dissolved oxygen concentrations were compared to more than 60 depth profiles at each of the three stations. Simulated results followed the same general patterns and magnitudes as measured values (fig. 22). The onset of low dissolved-oxygen concentrations and the recovery to higher dissolved-oxygen concentrations were well simulated throughout the reservoir. As with water temperature, occasionally, the depth of the simulated thermocline was a little above or below the measured thermocline providing large differences between simulated and measured dissolved oxygen concentrations within the layer of greatest change in temperature.

Simulated dissolved-oxygen concentrations ranged from 0.10 to 14.1 mg/L (fig. 22). Mean differences between measured and simulated dissolved-oxygen concentrations ranged from 0.06 mg/L to 0.78 mg/L with a RMSE that ranged from 1.35 to 1.50 mg/L and MAE that ranged from 0.89 to 1.18 mg/L for all stations over the simulation period (2004–10) (table 13).

Table 13. CE-QUAL-W2 model calibration evaluation statistics for water temperature, dissolved oxygen, selected nutrients, total organic carbon, chlorophyll *a*, and Secchi disk for Lake Maumelle stations.

[Difference is simulated minus measured]

Station	Year	Minimum difference	Maximum difference	Mean difference	Mean absolute error	Root mean square error
Temperature, in degrees Fahrenheit						
East of Highway 10	2004–10	-9.10	2.88	-1.07	1.53	2.09
Little Italy	2004–10	-0.71	6.46	-0.60	1.26	1.69
Natural steps	2004–10	-6.51	4.98	-0.49	1.31	1.68
Dissolved oxygen, in milligrams per liter						
East of Highway 10	2004–10	-2.83	5.52	0.78	1.18	1.50
Little Italy	2004–10	-0.74	5.03	0.06	0.89	1.39
Natural steps	2004–10	-7.80	4.70	0.20	0.90	1.35
Dissolved nitrite plus nitrate, in milligrams per liter as nitrogen						
East of Highway 10	2004–10	-0.133	0.047	-0.010	0.019	0.030
Little Italy	2004–10	-0.063	0.132	0.022	0.033	0.047
Natural steps	2004–10	-0.068	0.113	0.023	0.034	0.046
Dissolved ammonia, in milligrams per liter as nitrogen						
East of Highway 10	2004–10	-0.068	0.143	0.005	0.021	0.029
Little Italy	2004–10	-0.196	0.388	0.040	0.053	0.088
Natural steps	2004–10	-0.258	0.312	0.035	0.053	0.085
Total nitrogen, in milligrams per liter as nitrogen						
East of Highway 10	2004–10	-0.517	0.030	-0.124	0.126	0.152
Little Italy	2004–10	-0.330	0.350	-0.049	0.082	0.105
Natural steps	2004–10	-0.490	0.204	-0.045	0.084	0.112
Dissolved orthophosphate, in milligrams per liter as phosphorus						
East of Highway 10	2004–10	-0.004	0.013	0.000	0.002	0.003
Little Italy	2004–10	-0.005	0.034	0.005	0.007	0.011
Natural steps	2004–10	-0.024	0.031	0.002	0.005	0.008
Total phosphorus, in milligrams per liter as phosphorus						
East of Highway 10	2004–10	-0.041	0.013	-0.007	0.009	0.011
Little Italy	2004–10	-0.020	0.040	0.003	0.006	0.009
Natural steps	2004–10	-0.055	0.039	0.001	0.008	0.013
Total organic carbon, in milligrams per liter as carbon						
East of Highway 10	2004–10	-6.10	4.54	-0.55	1.05	1.41
Little Italy	2004–10	-3.71	0.91	-0.71	0.82	1.11
Natural steps	2004–10	-3.63	1.02	-0.70	0.83	1.17
Chlorophyll *a*, in micrograms per liter						
East of Highway 10	2004–10	-8.74	3.14	-0.62	1.73	2.49
Little Italy	2004–10	-6.73	3.36	-0.49	1.92	2.57
Natural steps	2004–10	-6.75	3.65	-0.30	1.96	2.65
Secchi disk depth, in feet						
East of Highway 10	2004–10	-2.6	5.4	1.6	1.9	2.3
Little Italy	2004–10	-3.9	3.7	0.6	1.4	1.8
Natural steps	2004–10	-6.2	5.1	0.3	1.6	2.1

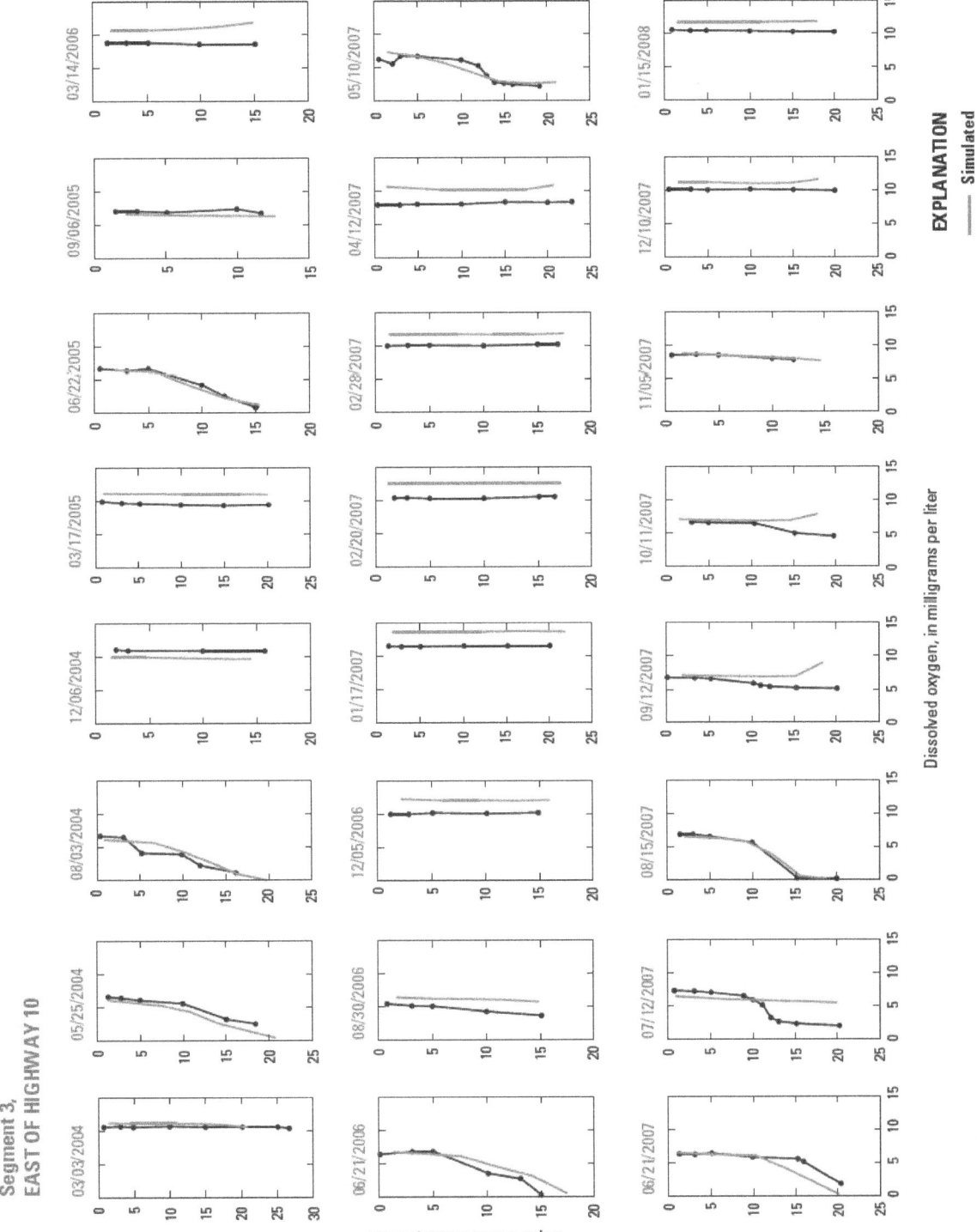

Figure 22.   Simulated and measured dissolved oxygen-depth profiles for Lake Maumelle.

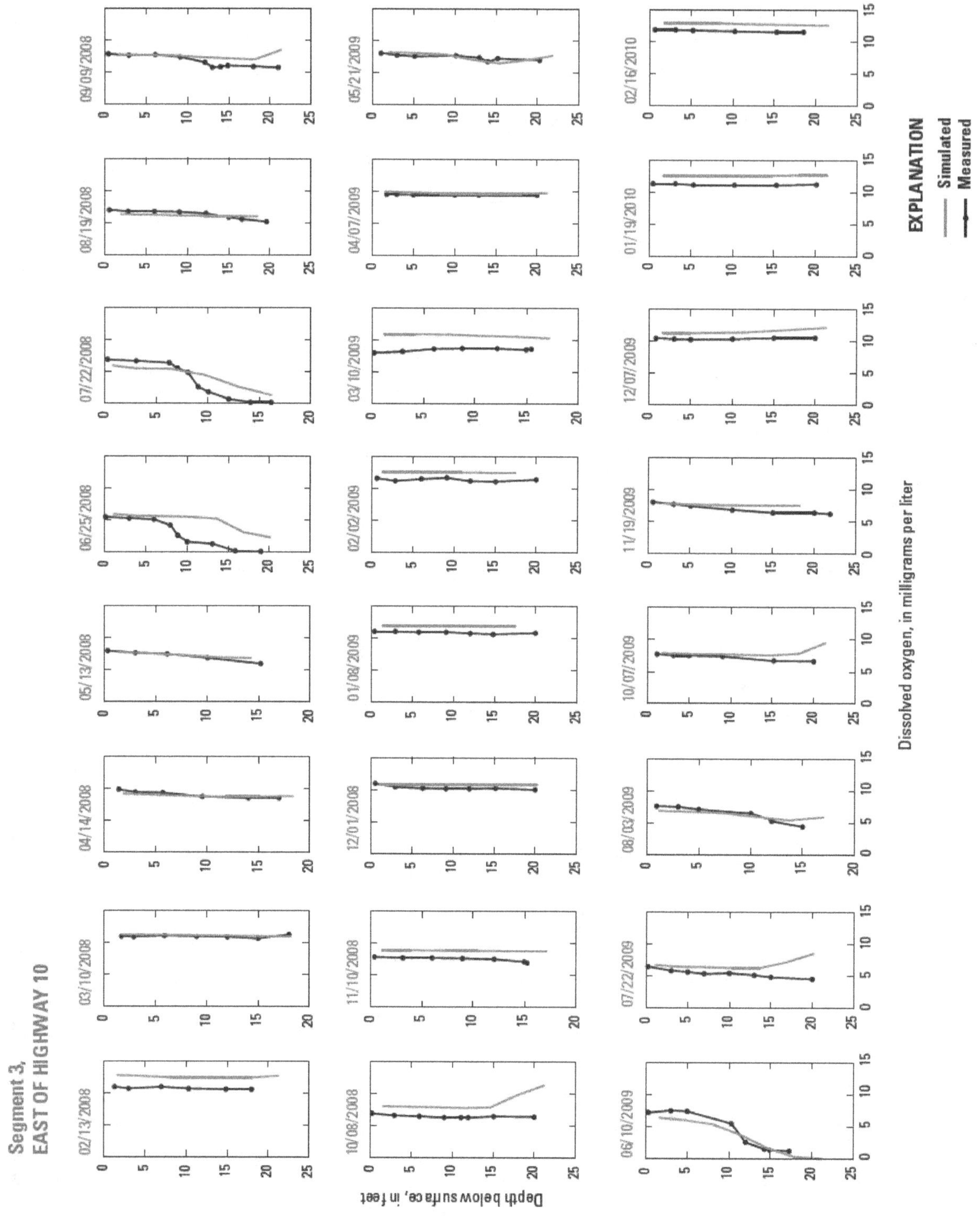

Segment 3,
EAST OF HIGHWAY 10

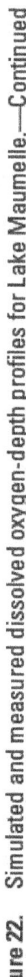

Dissolved oxgyen, in milligrams per liter

Depth below surface, in feet

EXPLANATION
— Simulated
—●— Measured

Figure 22.   Simulated and measured dissolved oxygen-depth profiles for Lake Maumelle.—Continued

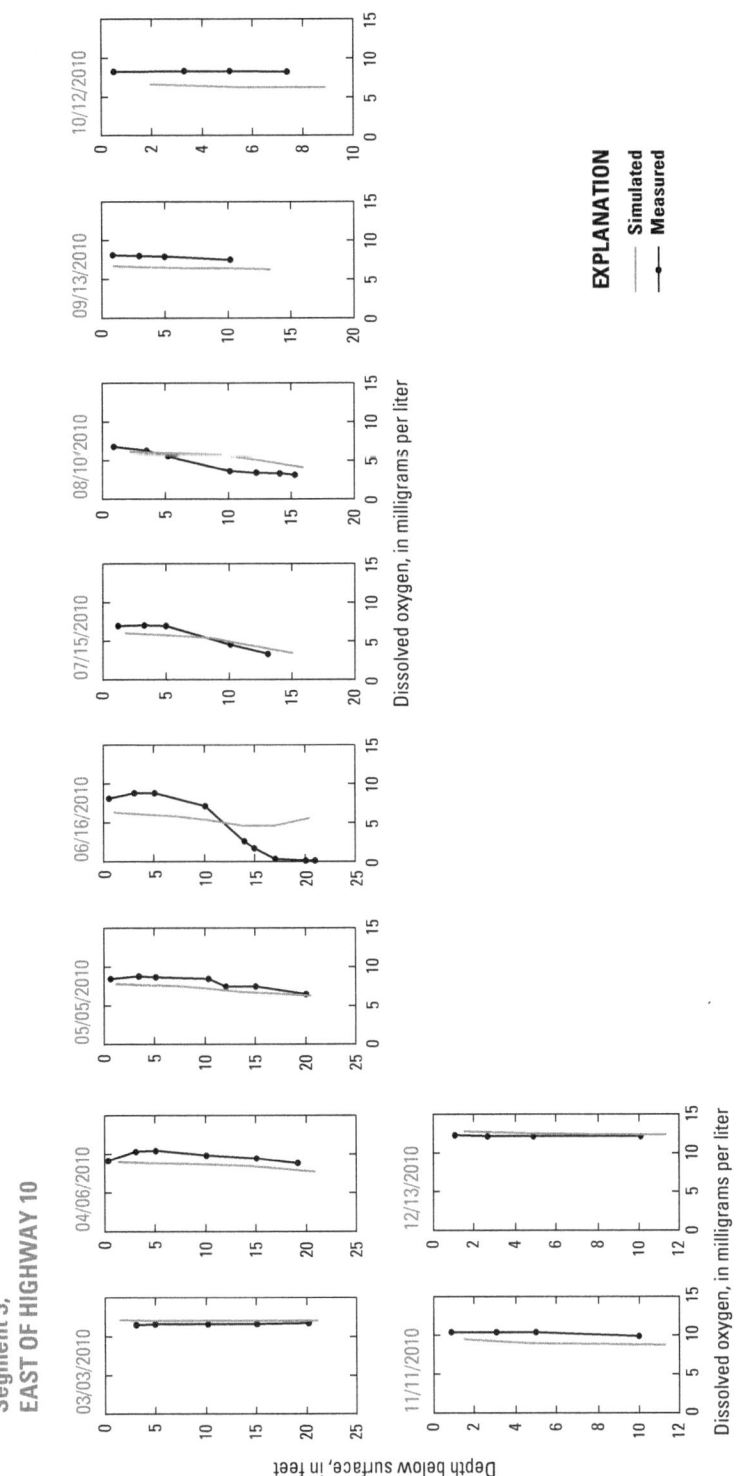

Figure 22. Simulated and measured dissolved oxygen-depth profiles for Lake Maumelle.—Continued

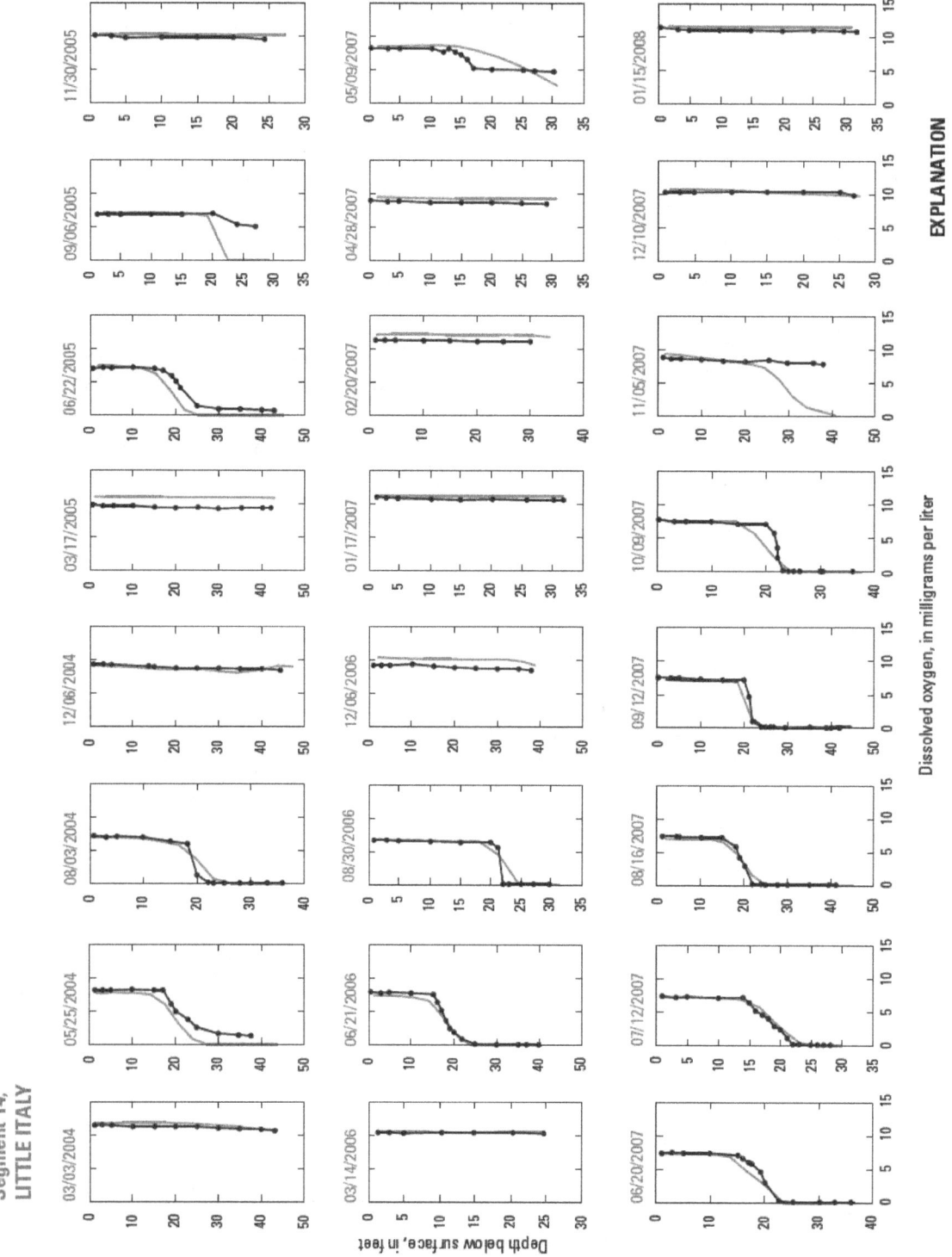

Figure 22. Simulated and measured dissolved oxygen-depth profiles for Lake Maumelle.—Continued

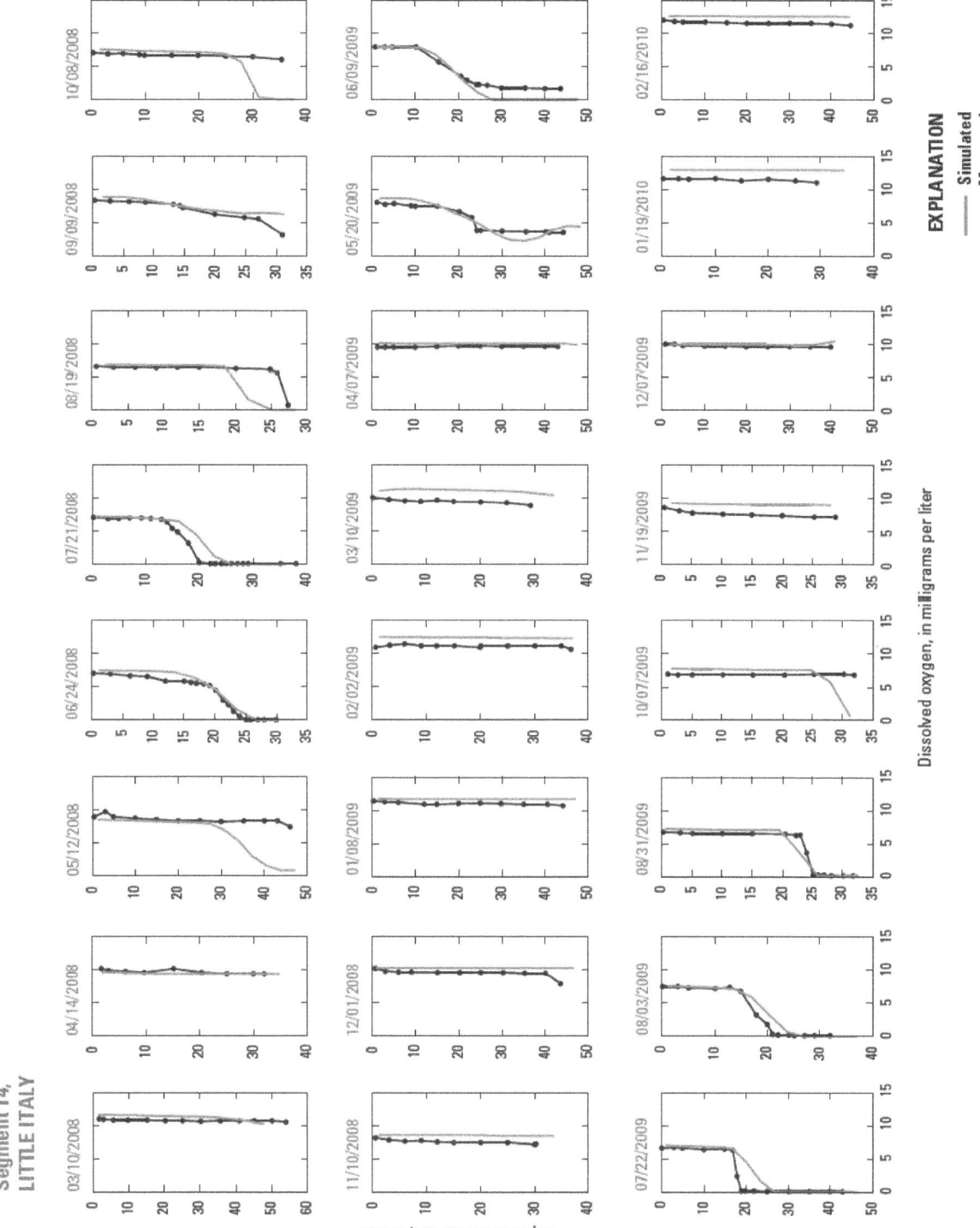

Figure 22.   Simulated and measured dissolved oxygen-depth profiles for Lake Maumelle.—Continued

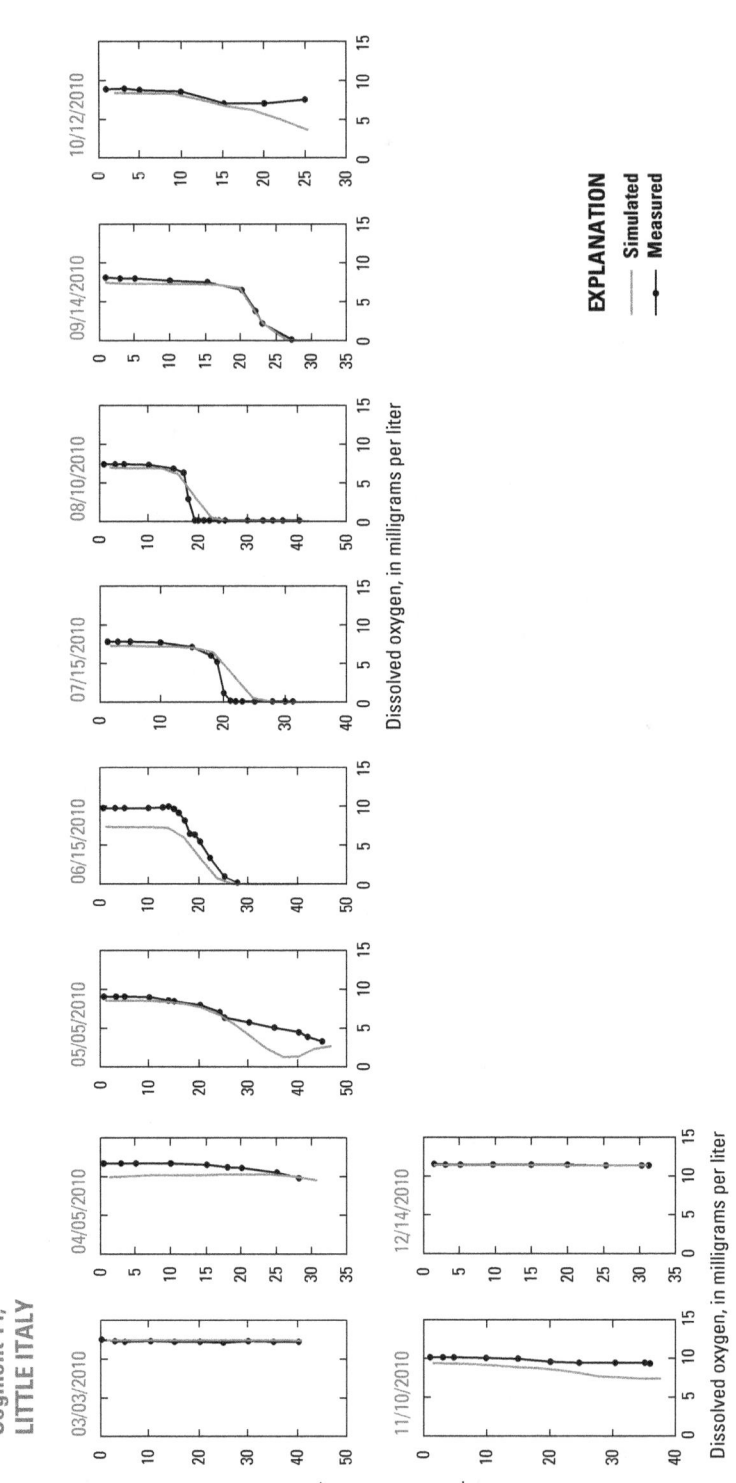

Figure 22.   Simulated and measured dissolved oxygen-depth profiles for Lake Maumelle.—Continued

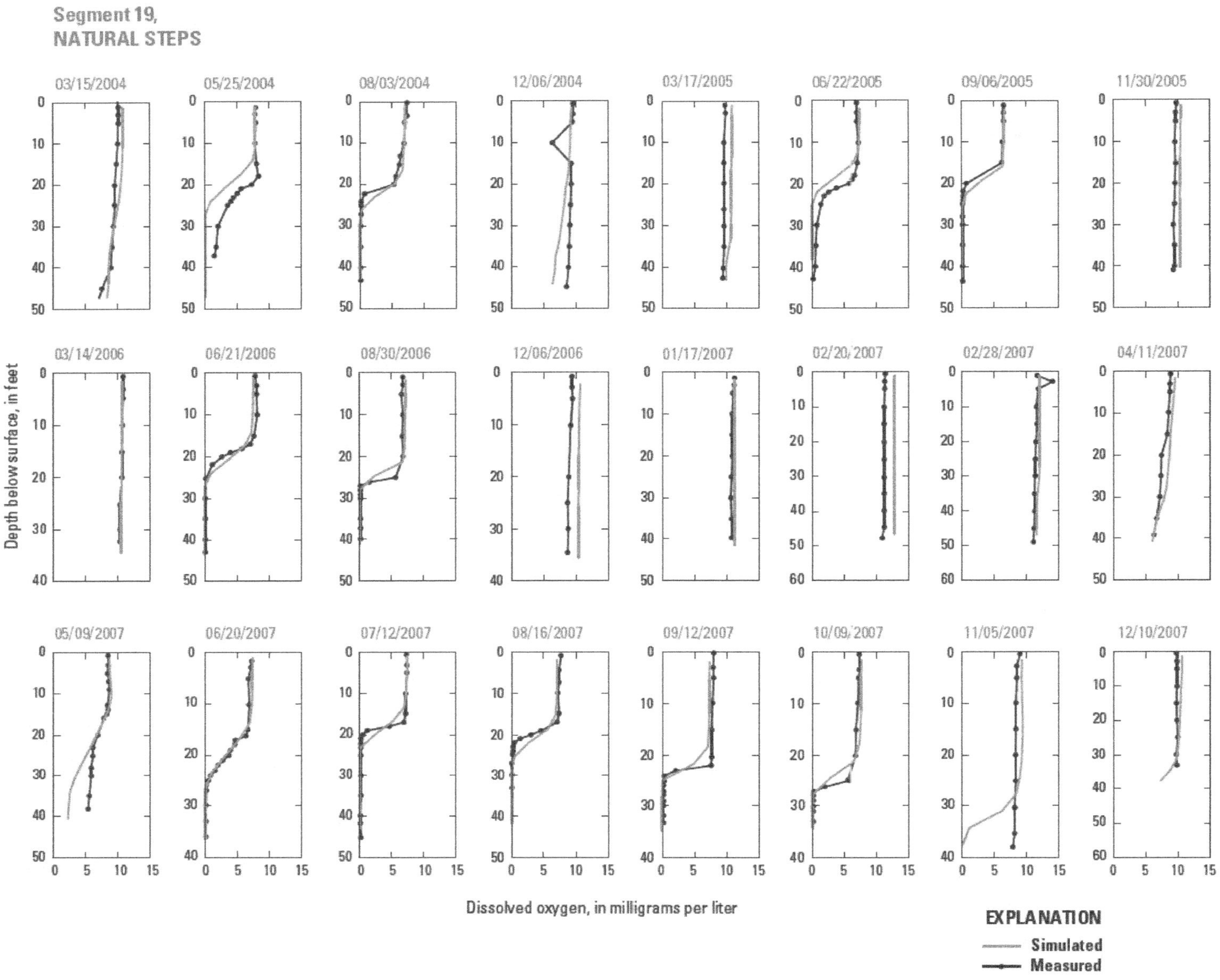

Segment 19, NATURAL STEPS

EXPLANATION

Simulated
Measured

Dissolved oxygen, in milligrams per liter

Depth below surface, in feet

Figure 22. Simulated and measured dissolved oxygen-depth profiles for Lake Maumelle.—Continued

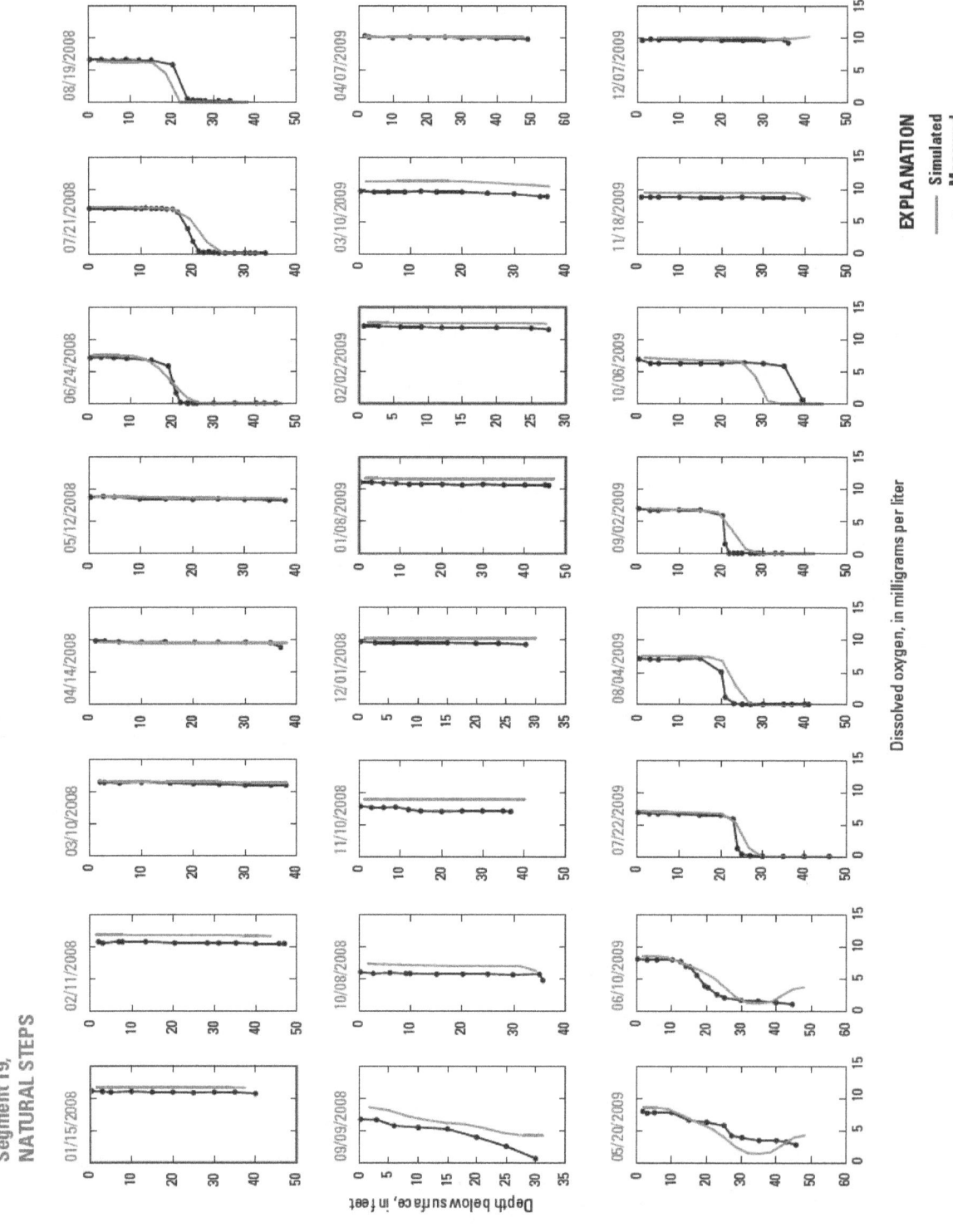

Figure 22. Simulated and measured dissolved oxygen-depth profiles for Lake Maumelle.—Continued

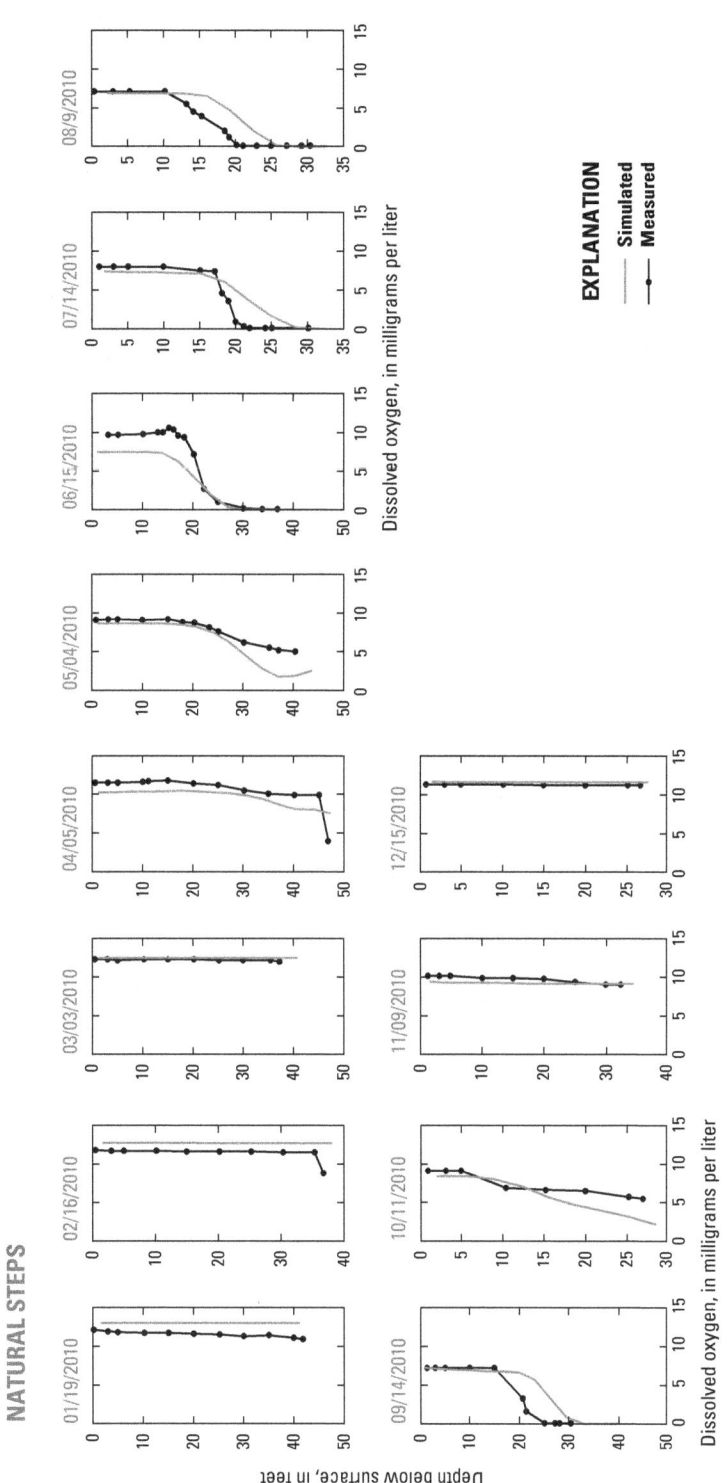

Figure 22. Simulated and measured dissolved oxygen-depth profiles for Lake Maumelle.—Continued

Dissolved Nitrite plus Nitrate Nitrogen, Dissolved Ammonia Nitrogen, Total Nitrogen, Dissolved Orthophosphate, and Total Phosphorus

Considering the oligotrophic-mesotrophic (low to intermediate primary productivity and associated low nutrient concentrations) condition of Lake Maumelle (Green, 1994), simulated algae (phytoplankton, chlorophyll *a*), phosphorus, and nitrogen concentrations compared well with generally low measured values. Often, simulated values were slightly lower than measured values. As discussed earlier in the "Previous Investigations" section, measured phosphorus and ammonia concentrations in Lake Maumelle were one to two orders of magnitude lower than estimates of national background concentrations (Green, 2001). The temporal changes in the concentrations of nitrogen and phosphorus in the reservoir in the Lake Maumelle model also were simulated reasonably well (figs. 23–24).

Simulated dissolved nitrite plus nitrate nitrogen concentrations in the epilimnion and the hypolimnion at reservoir stations Little Italy and Natural Steps, in general, matched measured concentrations (fig. 23). Measured and simulated dissolved nitrite plus nitrate nitrogen concentrations within the hypolimnion of the reservoir generally were higher than concentrations in the epilimnion. Dissolved nitrite plus nitrate nitrogen concentrations tended to be lower at the downstream stations during the summer stratification period (typically May to September) and increased slightly following turnover during the winter nonstratification period, a time when the reservoir is approximately uniform in temperature and mixing occurs throughout the entire water column. Mean differences between measured and simulated dissolved nitrite plus nitrate nitrogen concentrations ranged from -0.010 to 0.023 mg/L with a RMSE that ranged from 0.030 to 0.047 mg/L and a MAE that ranged from 0.019 to 0.034 mg/L for all stations over the simulation period (2004–10) (table 13).

Simulated dissolved ammonia nitrogen concentrations in the epilimnion for all three reservoir stations, in general, matched the measured concentrations, which were consistently near the LRL of 0.02 mg/L (fig. 23). Measured and simulated dissolved ammonia nitrogen concentrations within the hypolimnion at reservoir stations Little Italy and Natural Steps were higher than concentrations in the epilimnion. Dissolved ammonia nitrogen commonly accumulates in the hypolimnion layer in conjunction with the onset of anoxic conditions. Dissolved ammonia nitrogen can build up on sediments as organic matter decomposes at the bottom and then be released into the water column when the lake mixes. However, when the water is oxygenated, generally when the lake is not stratified, sediment release of dissolved ammonia nitrogen is negligible. The measured dissolved ammonia nitrogen concentrations in the hypolimnion for the two lake stations of Little Italy and Natural Steps began to increase as the temperatures began to warm during May and June.

The highest dissolved ammonia nitrogen concentrations were measured during July through September. Furthermore, measured dissolved ammonia nitrogen concentrations began to decline as the lake began to mix again with the onset of cooler temperatures. The simulated dissolved ammonia nitrogen concentrations matched this seasonal trend of increasing and decreasing dissolved ammonia nitrogen concentrations associated with stratification. Accordingly, the simulated and measured dissolved ammonia nitrogen concentrations for station East of Highway 10 for both the epilimnion and hypolimnion were low, near the LRL. This station is shallower and influenced by the inflow of Maumelle River that allows it to be mixed more than the other two lake stations; therefore, the dissolved ammonia nitrogen concentrations for the hypolimnion were similar to the concentrations within the epilimnion. Mean differences between measured and simulated dissolved ammonia nitrogen concentrations ranged from 0.005 mg/L to 0.040 mg/L with a RMSE that ranged from 0.029 to 0.088 mg/L and a MAE that ranged from 0.021 to 0.053 mg/L for all stations over the simulation period (2004–10) (table 13).

Simulated total nitrogen concentrations in the epilimnion and the hypolimnion at reservoir stations Little Italy and Natural Steps matched measured concentrations the first half of the period of record and dropped below measured concentrations after 2006 (fig. 23). Measured and simulated total nitrogen concentrations within the hypolimnion of the reservoir were higher than concentrations in the epilimnion. Mean differences between measured and simulated total nitrogen concentrations ranged from -0.124 to -0.045 mg/L with a RMSE that ranged from 0.105 to 0.152 mg/L and a MAE that ranged from 0.082 to 0.126 mg/L for all stations over the simulation period (2004–10) (table 13).

Simulated dissolved orthophosphate concentrations in the epilimnion and the hypolimnion at reservoir stations Little Italy and Natural Steps compared well with measured concentrations (fig. 24). Mean differences between measured and simulated dissolved orthophosphate concentrations ranged from 0.000 to 0.005 mg/L with a RMSE that ranged from 0.003 to 0.011 mg/L and a MAE that ranged from 0.002 to 0.007 mg/L for all stations over the simulation period (2004–10) (table 13).

Simulated total phosphorus concentrations in the epilimnion and the hypolimnion at reservoir stations Little Italy and Natural Steps generally matched measured concentrations (fig. 24). Measured and simulated total phosphorus concentrations within the hypolimnion of the reservoir often were higher than concentrations in the epilimnion. The high measured concentrations in the hypolimnion at Natural Steps could not be simulated because often these samples were collected 3.28 ft (1 m) above the bottom inside the old river channel. The model geometry was not resolute enough to geometrically define the thin, deep river channel at the downstream end of the reservoir.

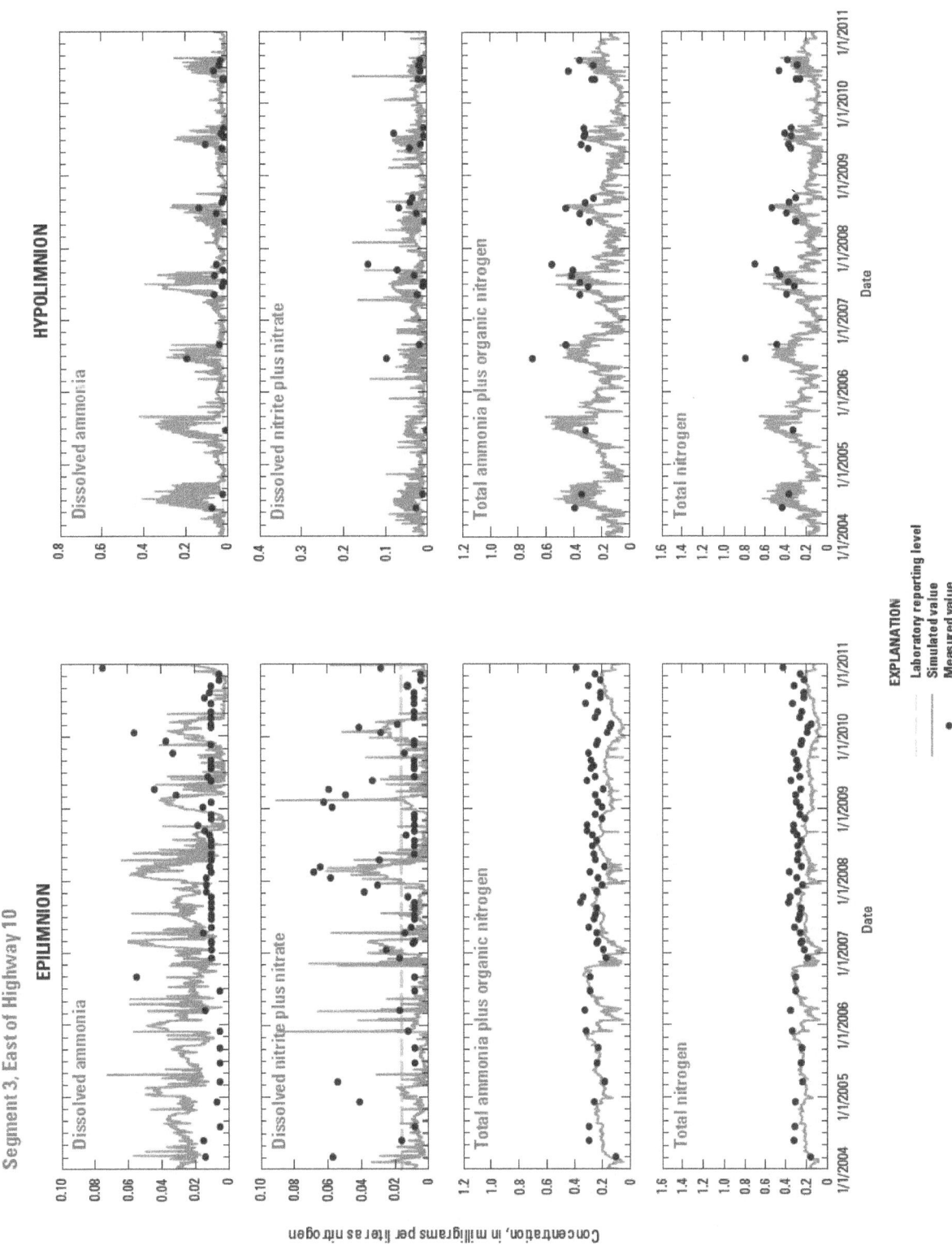

Figure 23.   Simulated and measured nitrogen concentrations in Lake Maumelle.

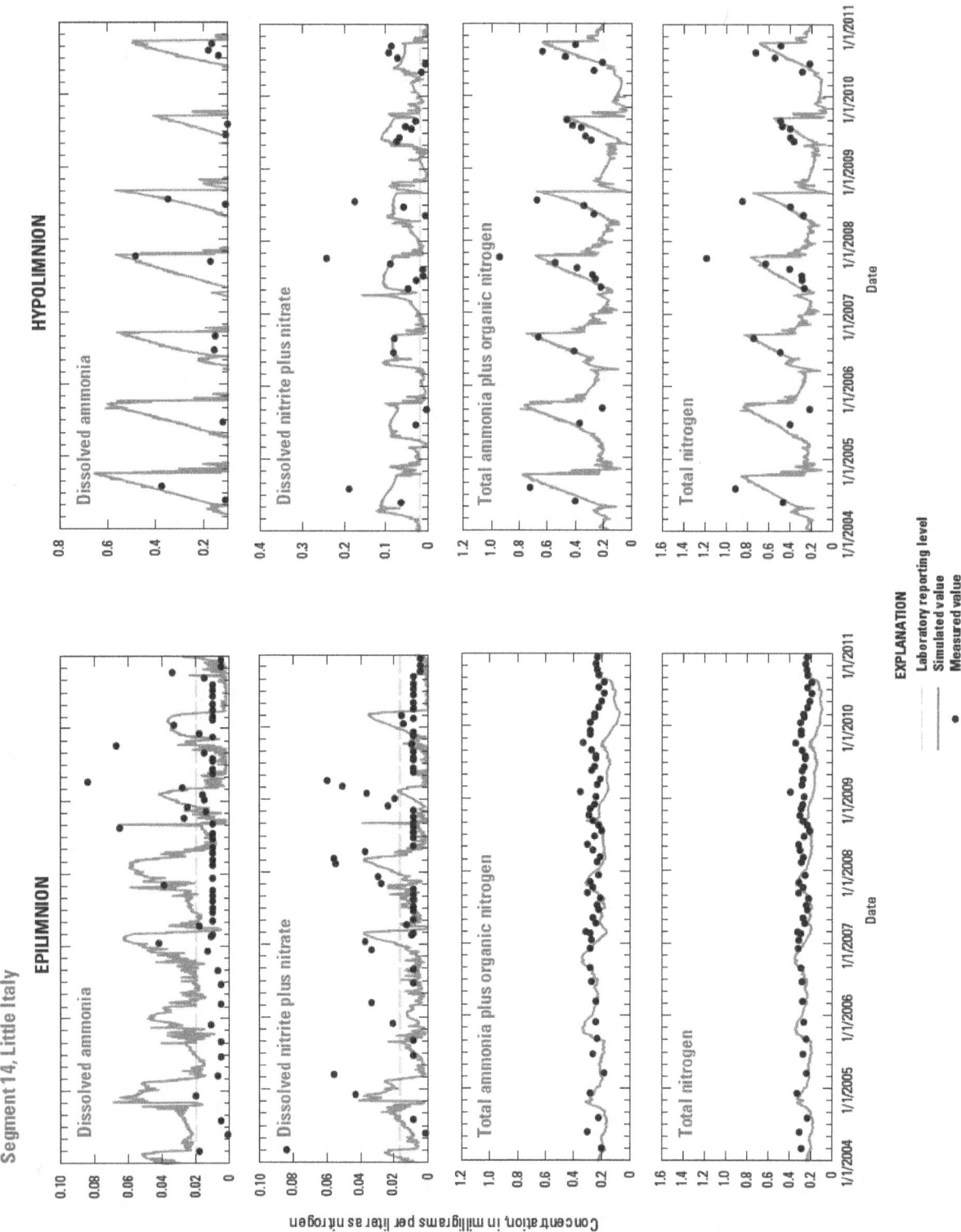

Figure 23.   Simulated and measured nitrogen concentrations in Lake Maumelle.—Continued

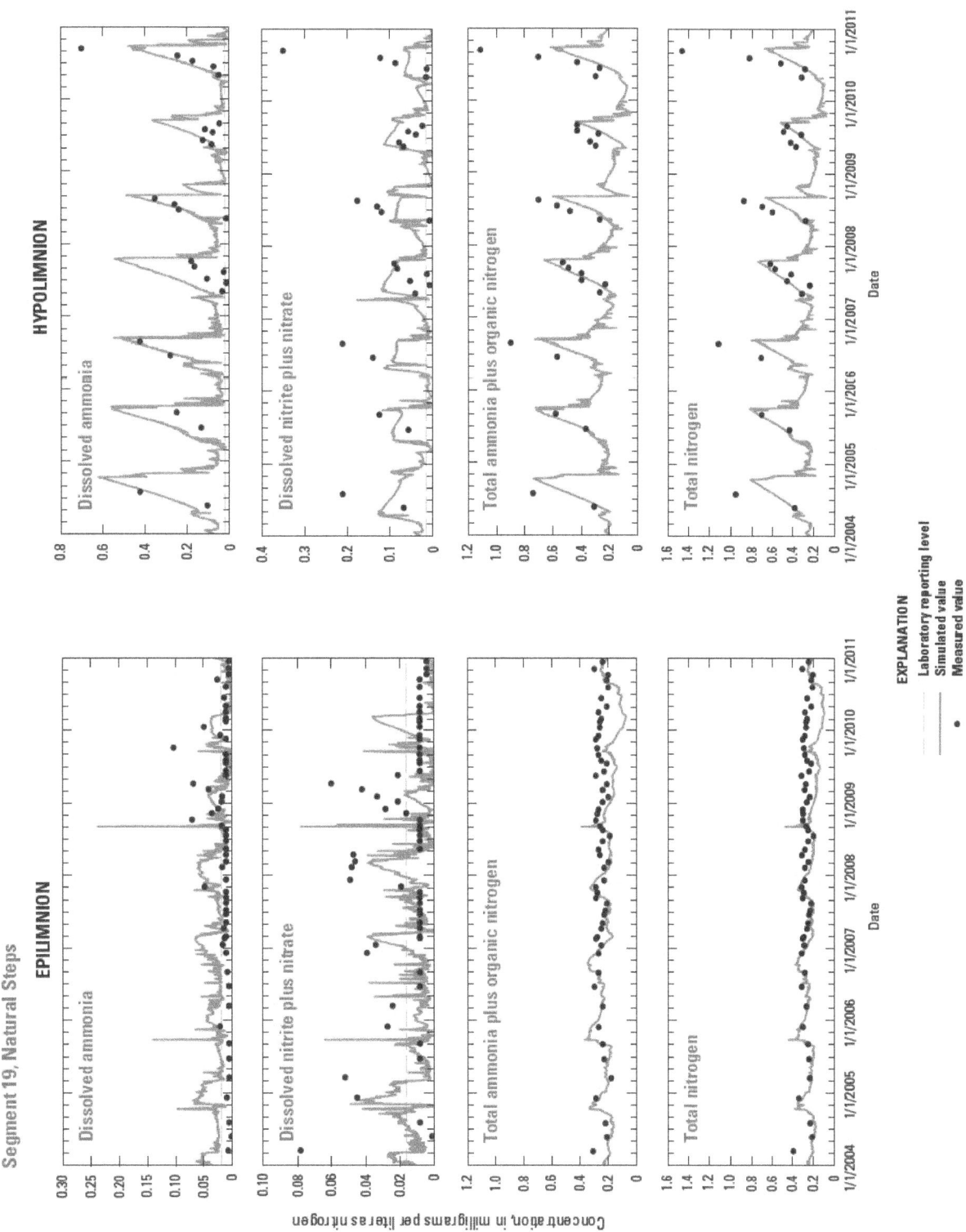

Figure 23. Simulated and measured nitrogen concentrations in Lake Maumelle.—Continued

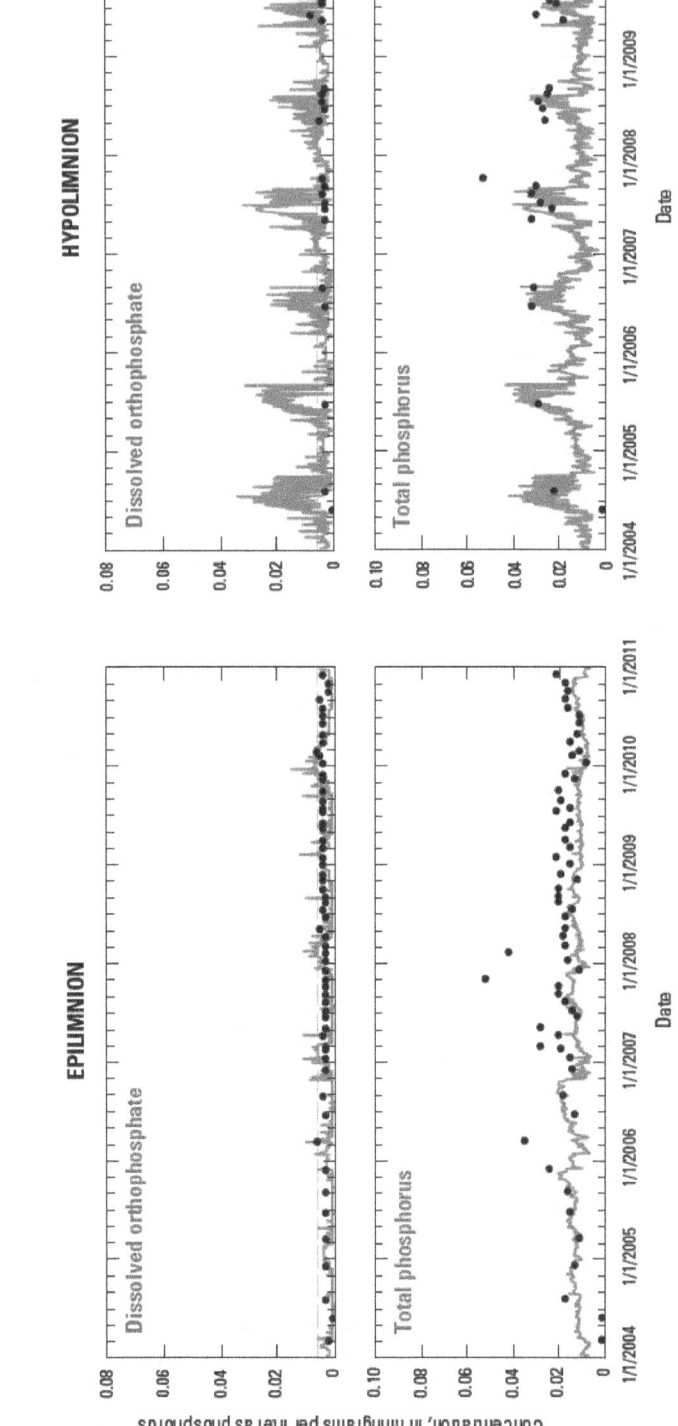

Figure 24.   Simulated and measured phosphorus concentrations in Lake Maumelle.

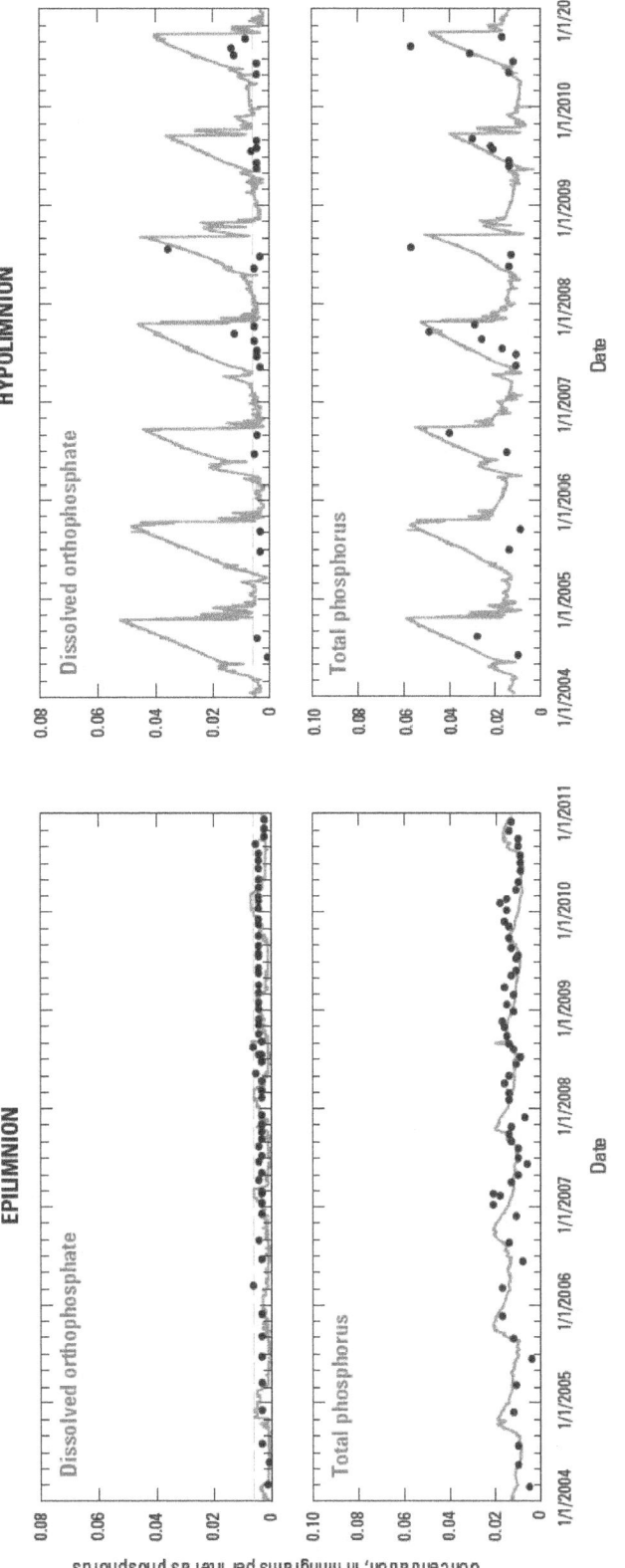

Figure 24. Simulated and measured phosphorus concentrations in Lake Maumelle.—Continued

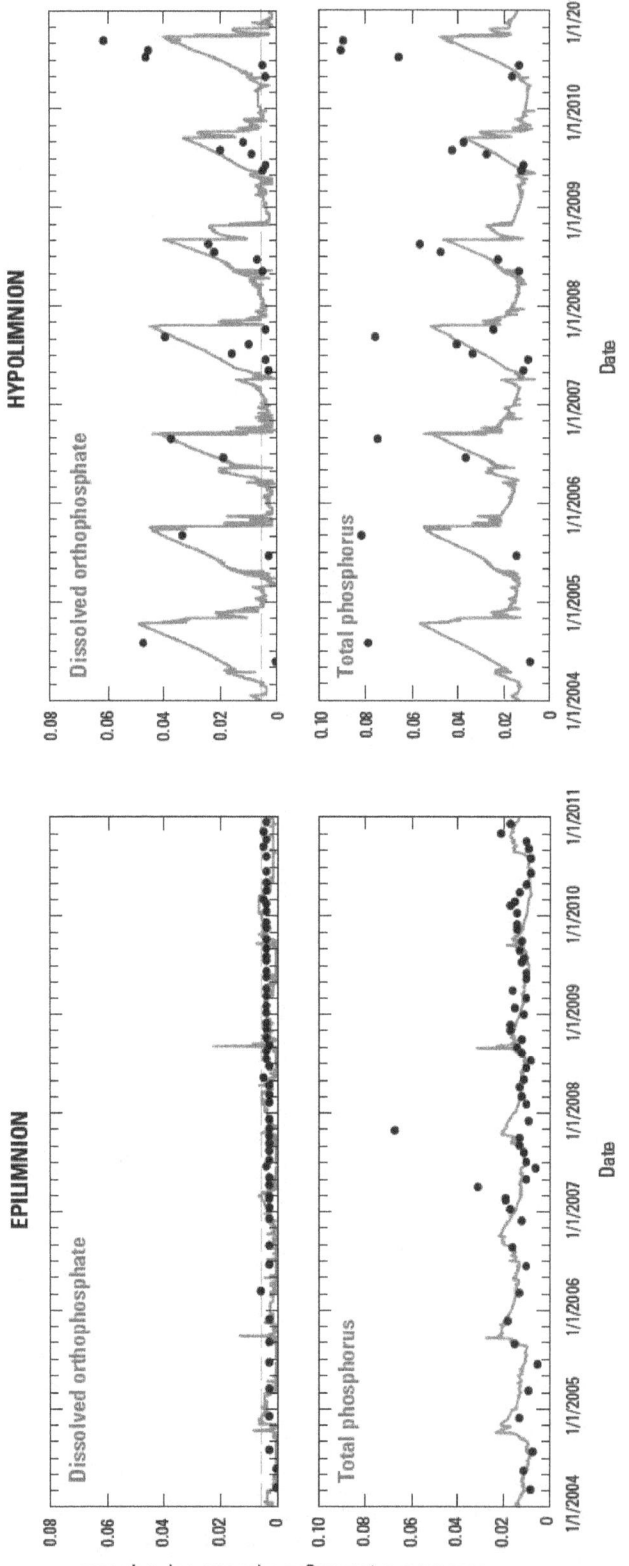

Figure 24.   Simulated and measured phosphorus concentrations in Lake Maumelle.—Continued

Simulated total phosphorus concentrations tended to be lower at the downstream stations during the summer stratification period (typically May to September) and increased slightly following turnover during the winter nonstratification period, a time when the reservoir is approximately uniform in temperature and mixing throughout the entire water column. Mean differences between measured and simulated total phosphorus concentrations were -0.007 to 0.003 mg/L with a RMSE that ranged from 0.009 to 0.013 mg/L and a MAE that ranged from 0.006 to 0.009 mg/L for all stations over the simulation period (2004–10) (table 13).

Algae

Simulated algal biomass converted to chlorophyll *a* was compared to measured chlorophyll *a* concentrations at the three stations in Lake Maumelle. The concentration of chlorophyll *a* commonly is used as a measure of the density of the algal population of a reservoir. The CE–QUAL–W2 model converts simulated algal biomass into chlorophyll *a* using a chlorophyll-algae ratio (table 4), which can be compared to measured data. Some of the differences between simulated and measured data can be explained by the variability in the measured chlorophyll *a* concentrations because of the nonhomogeneous vertical distribution of phytoplankton in the water column. However, the simulated concentrations of chlorophyll *a* followed the pattern of occurrence in the measured data with greater concentrations usually occurring from July through November (peaking following autumn mixing), and lower concentrations usually occurring from December through June (fig. 25). Mean differences between measured and simulated chlorophyll *a* concentrations ranged from -0.62 to -0.30 µg/L with a RMSE that ranged from 2.49 to 2.65 µg/L and a MAE that ranged from 1.73 to 1.96 µg/L for all stations over the simulation period (2004–10) (table 13).

Secchi disk

Secchi disk depths measure water clarity; the greater the depth value the clearer the water will be in the upper part of the water column. Water clarity is associated with the amount of suspended material within the water column and, therefore, can be used to evaluate water quality. Daily Secchi disk values were derived from selected CE–QUAL–W2 output and compared to measured values at the three sampling locations within Lake Maumelle. The CE–QUAL–W2 model calculates the light extinction coefficient in each layer of water over time. A constant of 1.7 was then divided by the extinction coefficient of the top layer of water to derive Secchi disk depth.

The derived Secchi disk measurements for Lake Maumelle exhibited a seasonal trend of greater water clarity (deeper depths) during the summer months and lower water clarity (shallower depths) during the fall months, generally following the same temporal trend as the measured Secchi disk depths (fig. 26). Water clarity was hardest to match at the station East of Highway 10, which had the lowest mean simulated and measured water clarity, 6.0 ft (n is 2,558) and 4.2 ft (n is 59), respectively, and the largest range in simulated Secchi disk depths (0.4 to 9.8 ft), when compared to the other two reservoir stations. The station at Little Italy had a mean measured depth of 6.4 ft (n is 59) and a mean simulated depth of 7.1 ft (n is 2,558) ranging from 3.6 to 9.5 ft. The station at Natural Steps had a mean measured depth of 6.7 ft (n is 60) and a mean simulated depth of 7.1 ft (n is 2,558) ranging from 2.8 to 9.5 ft (fig. 26). Mean differences between measured and simulated Secchi disk depths ranged from 0.3 to 1.6 ft with a RMSE that ranged from 1.8 to 2.3 ft and a MAE that ranged from 1.4 to 1.9 ft for all stations over the simulation period (2004–10) (table 13).

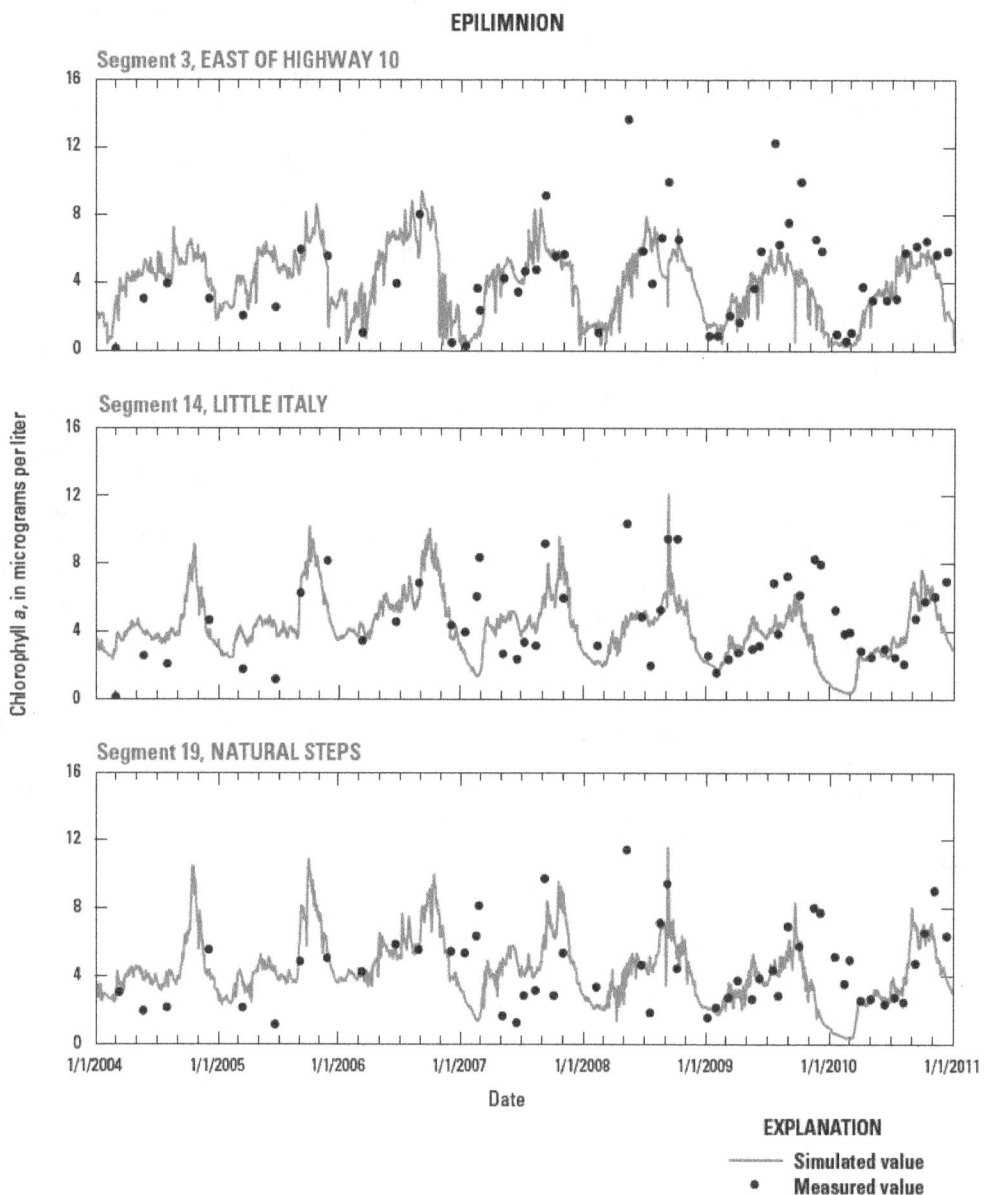

Figure 25.   Simulated and measured chlorophyll *a* concentrations in Lake Maumelle.

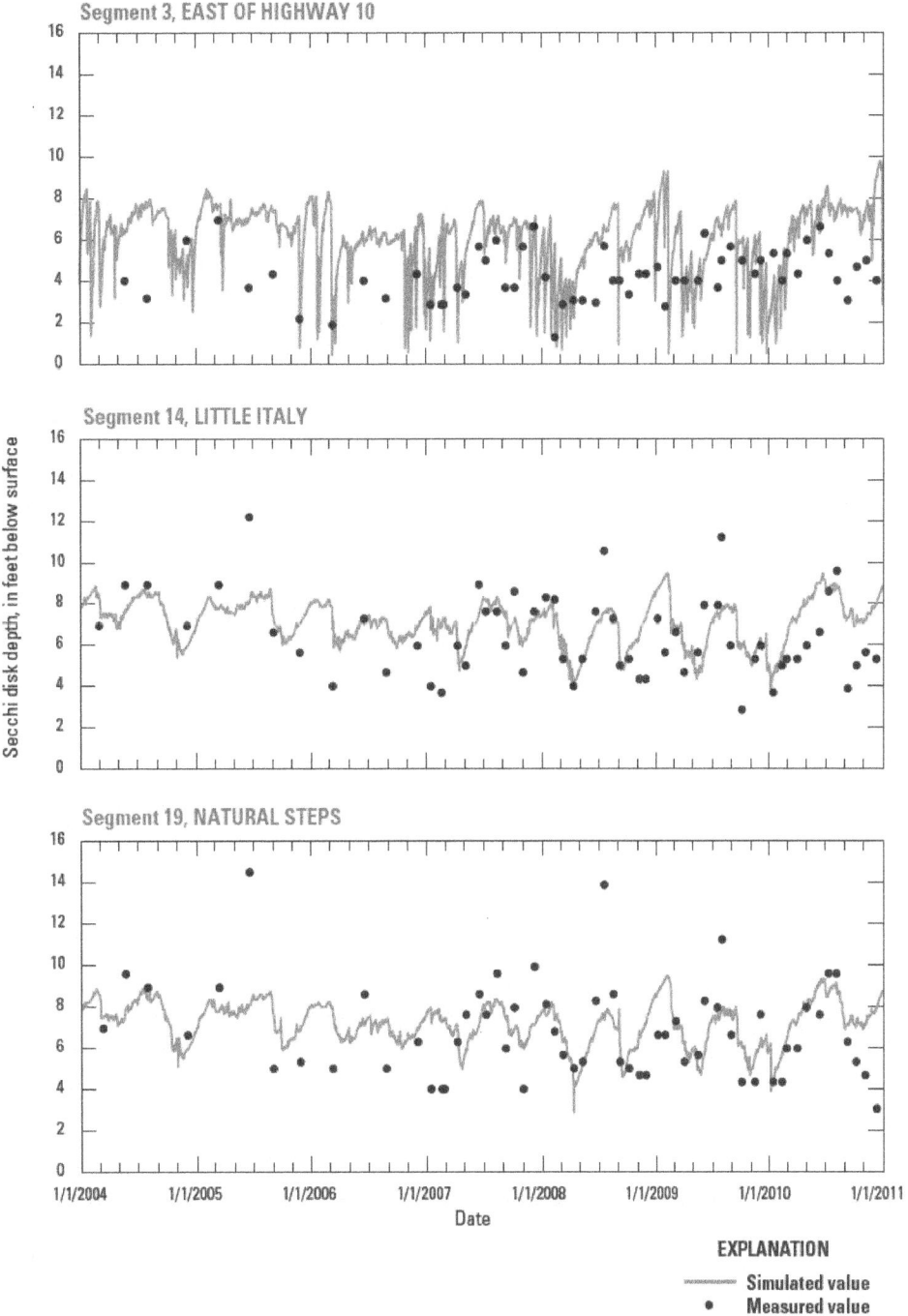

Figure 26. Derived and measured Secchi disk measurements in Lake Maumelle

Organic Carbon

Organic carbon concentrations simulated as dissolved and total organic carbon within the reservoir were relatively constant temporally in the reservoir, except at the upstream station, East of Highway 10. Spikes in measured concentrations, typically as a result of flood events, were difficult to simulate, but the CE–QUAL–W2 model simulates the general trend of increasing and decreasing organic carbon concentrations (fig. 27). Like total nitrogen, simulated organic carbon matched measured concentrations during 2004 through 2006 but consistently dropped lower than measured concentrations after 2006. Mean differences between measured and simulated total organic carbon concentrations ranged from -0.71 to -0.55 mg/L with a RMSE that ranged from 1.11 to 1.41 mg/L and a MAE that ranged from 0.82 to 1.05 mg/L for all stations over the simulation period (2004–10) (table 13).

Fecal Coliform Bacteria

Simulated fecal coliform concentrations exhibited the same general pattern and magnitudes as measured values (fig. 28). Higher fecal coliform densities occurred at the most upstream station and decreased downstream (fig. 28).

Hydrologic Simulation Program– FORTRAN and CE–QUAL–W2 Model Limitations

An understanding of model limitations is essential for the effective use and interpretation of watershed and reservoir models. The accuracies of the HSPF and CE–QUAL–W2 models are limited by the simplification of complexities with the physical properties of streamflow, hydrodynamics within the reservoir, water-quality processes within the watershed, by spatial and temporal discretization effects, and by assumptions made in the formulation of the governing equations. Model accuracy is limited by subwatershed and segment size, boundary conditions, accuracy of calibration, and parameter sensitivity; model accuracy also is limited by the availability of appropriate data and by the interpolations and extrapolations that are inherent in using data in any model. For example, streamflow and water quality can be simulated with reasonable accuracy at stream or reservoir locations with a streamflow or water-quality gaging station. However, at tributaries that do not have gaging stations, adequacy of streamflow simulation is uncertain. Although a model might be considered calibrated, calibration parameter values are not unique in yielding acceptable simulated values of streamflow, reservoir hydrodynamics, or water-quality characteristics.

Sensitivity analysis is the determination of the effects of small changes in calibrated model parameters on model results. A complete sensitivity analysis for all model parameters in the Lake Maumelle model was not conducted. The HSPF and CE–QUAL–W2 models include more than 150 parameters (tables 4, 7, and 9), and a complete sensitivity analysis would be a very lengthy process. However, many hydrodynamic and water-quality simulations were conducted as a component of model calibrations. Results from these simulations form the basis for the sensitivity analysis.

For the HSPF model, parameter sensitivity is a function of the physical conditions of Lake Maumelle's watershed, such as climate, topography, soils, and vegetation. For the simulation of streamflow, LZSN and INFILT (both of which play a direct role in determining whether moisture on the land surface infiltrates, enters storage, or becomes runoff) typically are the most sensitive hydrologic parameters (Al-Abed and Whiteley, 2002; Ryu, 2009; Skahill, 2003). Parameters related to detachment and washoff of sediment from the land surface, as well as the parameters related to shear stress within the stream reach, such as TAUCD and TAUCS, typically appeared to be among the most sensitive parameters for simulation of suspended sediment within the Lake Maumelle watershed HSPF model. Water temperature, water-quality constituents simulated as being sediment associated, and water-quality constituents that are removed simply by overland flow appeared most sensitive to KATRAD, POTFW and nutrient adsorption parameters, and WQSOP, respectively, in the Lake Maumelle watershed HSPF model.

HSPF is a one-dimensional model, and therefore, simulated concentrations can be erroneously high during periods of low flow because mass is conserved in sediment and nutrient simulations. However, because these anomalous simulated concentrations only occur during low flow periods, there is little or no volume of water to move these anomalously high concentrations, and the associated loads are very low. Therefore, the concentrations during these periods of low flow are not shown on the HSPF model time series figures (figs. 11–19) when streamflow was less than 1.8 ft³/s.

Furthermore, during storm events, it is difficult to be at each site to sample the peak flow, which presumably corresponds with the peak in suspended constituent load. As such, the model is simply limited by the paucity of data. To adequately predict and match simulated and observed values through entire stormflow events, a sampling program would have to be initiated to collect data along the entire stormflow hydrograph. Although positive linear relations will be found, for example, between streamflow and suspended sediment, there usually is large variation as a result of numerous variables. The resulting low R^2 values normally will reflect this variation as a result of sampling timing, difficulty in collecting representative samples, and varying landscape characteristics. As such, the model may actually perform much better as a predictive tool (especially for seasonal or annual loads rather than instantaneous or daily concentrations) than is reflected in attempts at matching individual simulations to observed instantaneous measurements.

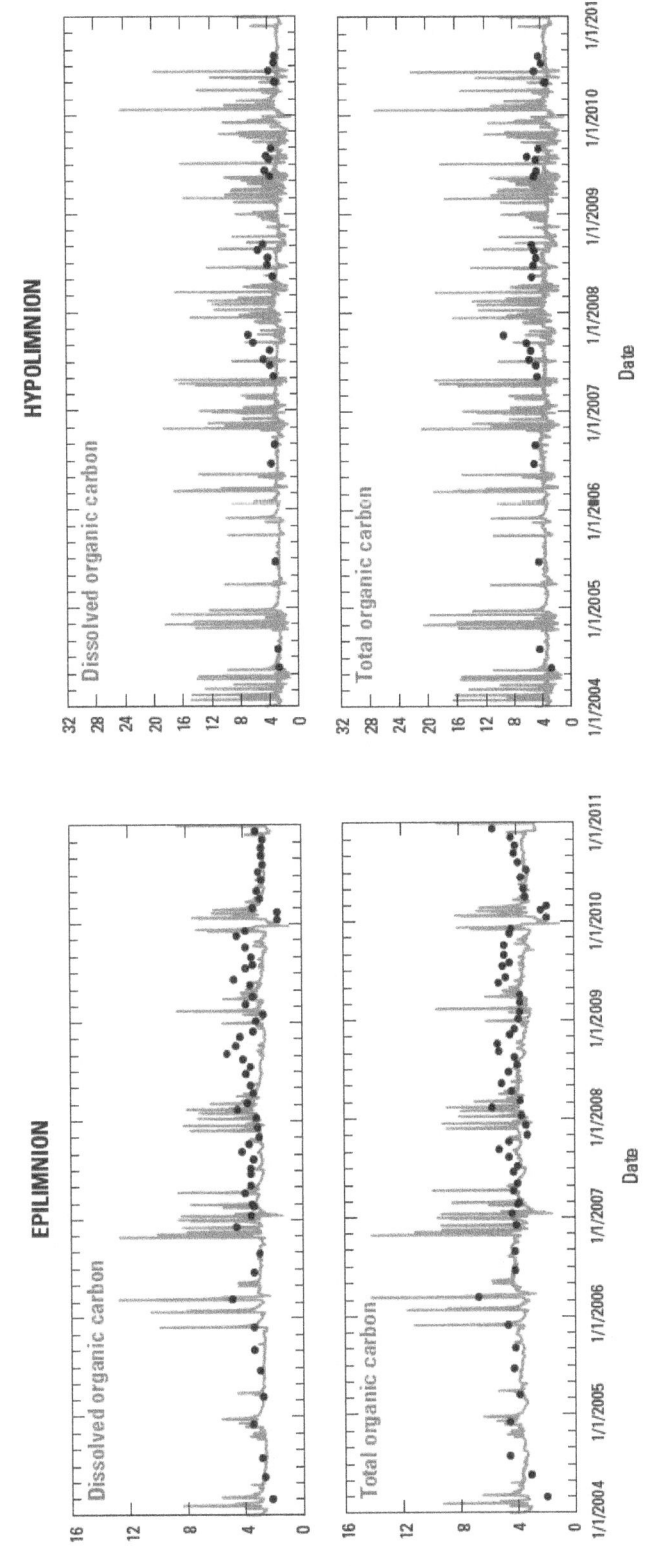

Figure 27.   Simulated and measured total organic carbon concentrations in Lake Maumelle.

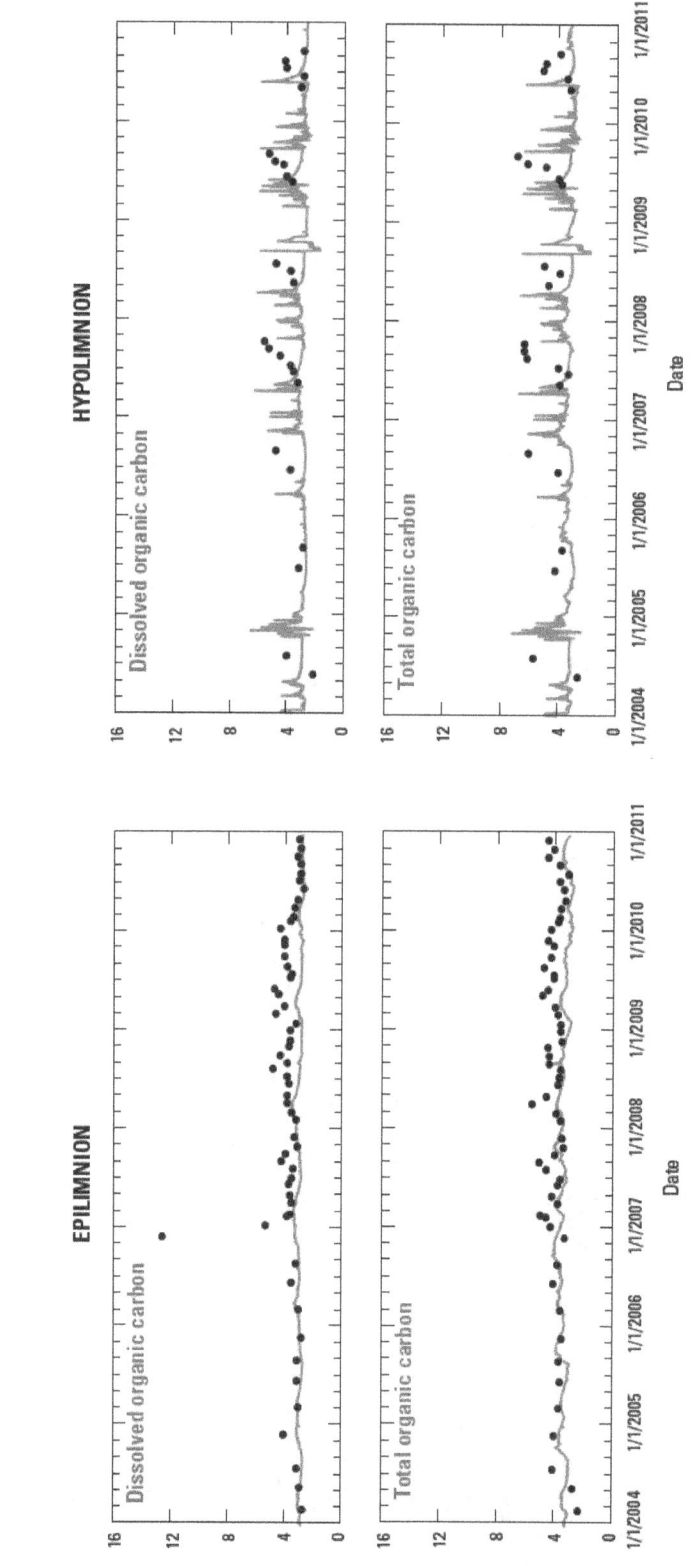

Figure 27.   Simulated and measured total organic carbon concentrations in Lake Maumelle.—Continued

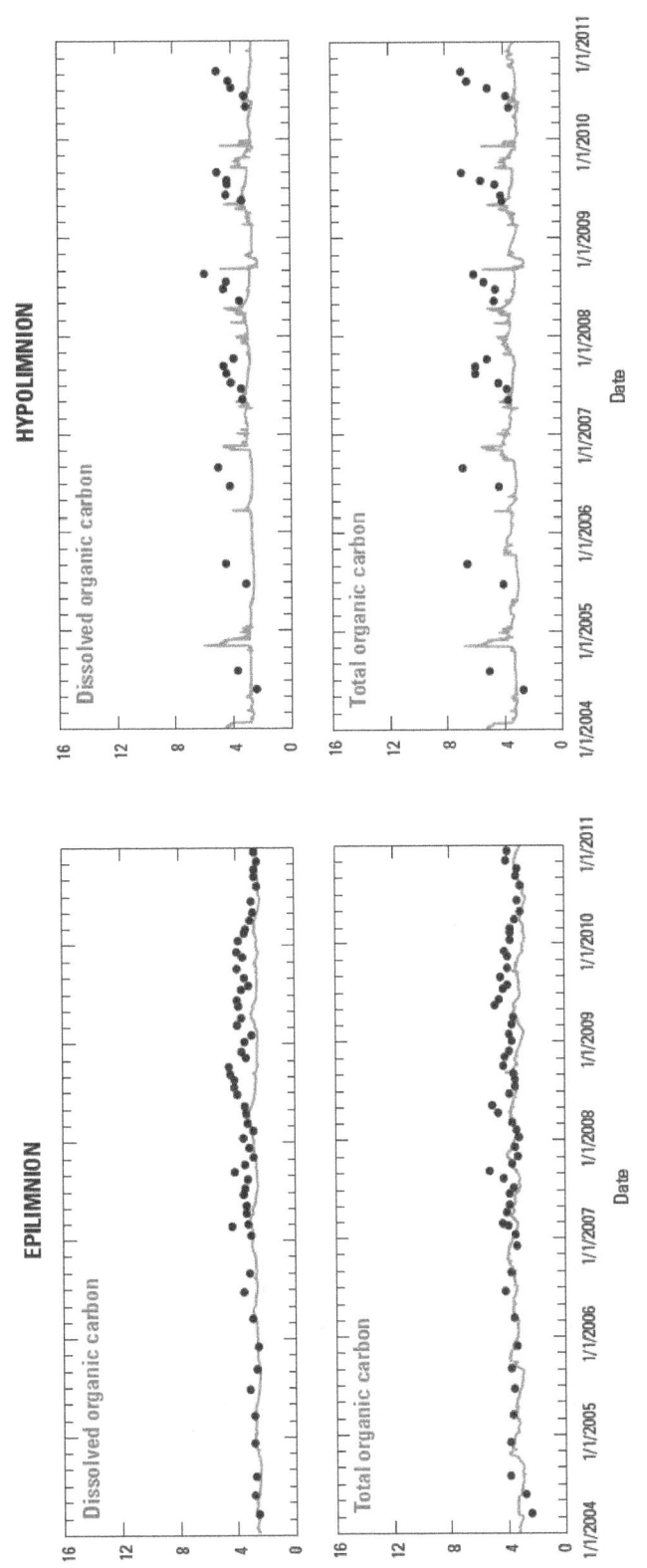

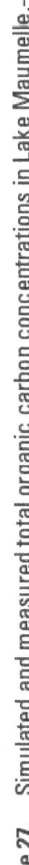

Figure 27.   Simulated and measured total organic carbon concentrations in Lake Maumelle.—Continued

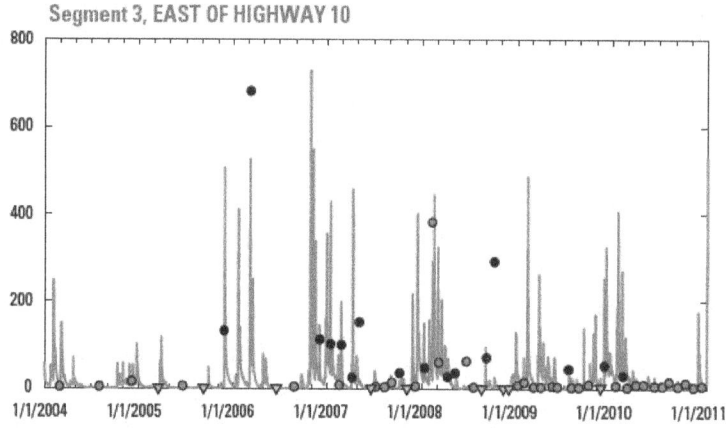

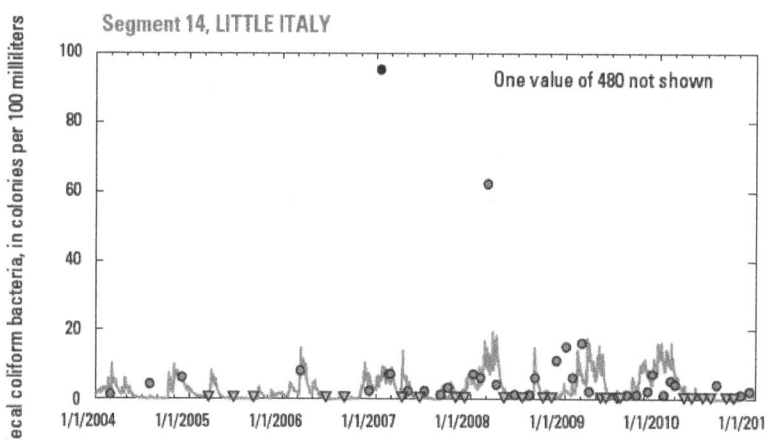

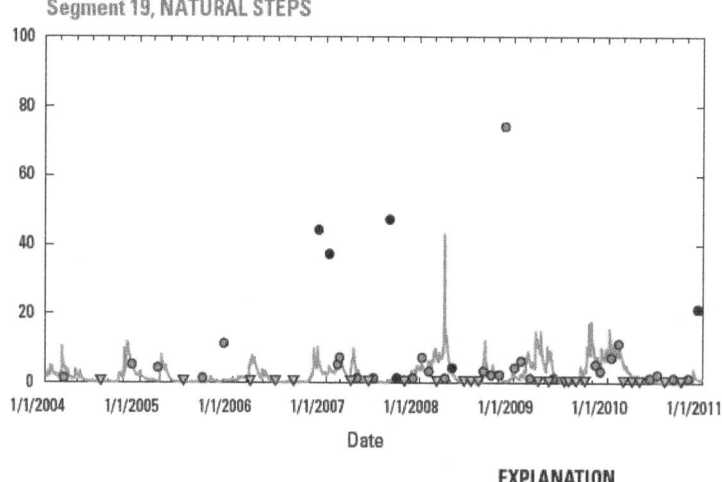

Figure 28.   Simulated and measured fecal coliform bacteria concentrations in Lake Maumelle.

An overriding variable affecting large discrepancies (as a percentage) between simulated and observed values is the low concentrations for many of the constituents. For example, simulated and observed ammonia nitrogen concentrations are actually less than the LRL for ammonia nitrogen analysis.

The quality of runoff water varies spatially and temporally—particularly temporally. Studies conducted by the University of Arkansas at Fayetteville (Haggard and others, 2005; Heathman and others, 1995; Kleinman and Sharpley, 2003; Sharpley, 1997) have shown that following application of manure to fields, the highest nutrient concentrations occur during the first runoff event, and concentrations approaching preapplication concentrations occur during the second and third runoff events. Additionally, in nearly pristine (dominantly forested) environments with little anthropogenic land use, runoff from early spring rains is enriched with nutrients and other organic matter (leaf litter, animal wastes, and other sources), with runoff from ensuing rains carrying lower nutrient and organic compound loads. As such, timing is important, and runoff events, no matter how large, would not necessarily carry similar nutrient and organic compound loads.

A limitation of the CE–QUAL–W2 model is that it is a two-dimensional representation of a three-dimensional waterbody. The governing equations are laterally and vertically averaged within layers. Although the model may accurately represent vertical and longitudinal processes within the reservoir, processes that occur laterally, or from shoreline to shoreline perpendicular to the downstream axis, may not be properly represented.

For the CE–QUAL–W2 model, simulated temperatures were most sensitive to changes in the wind-sheltering coefficient (WSC). Wind speed and direction were recorded at Little Rock Adams Field (fig. 5), about 18 mi southeast of Lake Maumelle Dam. Variability in wind speed and direction between Lake Maumelle and Adams Field can result in different mixing patterns in the model than what actually occurred in the lake, producing error and uncertainty in the depth of the mixing layer when thermally stratified and producing large differences in temperature for the same depth in the region of the thermocline. The coefficient of bottom heat exchange (CBHE) and light extinction coefficient (α) also affected position of the thermocline and temperatures above and below the thermocline. Simulated dissolved oxygen appeared to be most sensitive to changes in sediment oxygen demand and thermal patterns. Changes in the wind-sheltering coefficient, which affected thermal patterns, also affected dissolved oxygen concentration. Algal concentrations were most sensitive to the algal growth rate and were sensitive to mortality and settling rates. Algal growth during the cooler seasons was sensitive to light saturation, light intensity, and algal temperature-rate multipliers and also was affected by algal half saturation constants for phosphorus and nitrogen. The algal half saturation constants affected nutrient uptake and, therefore, nitrogen and phosphorus concentrations in the lake over time. Also, phosphorus and nitrogen concentrations were sensitive to release rates from bottom sediments.

Some limitations associated with the water-quality interactions in the CE–QUAL–W2 model include not simulating the effects of zooplankton and macrophytes and the model's use of simplistic sediment-oxygen demand computations. The zooplankton and macrophyte communities not represented in the CE–QUAL–W2 model may have an effect on how the phytoplankton community or recycling of nutrients are simulated. Eddy coefficients are used to model turbulence in a reservoir in which vertical turbulence equations are written in the conservative form using the Boussinesq and hydrostatic approximations (Cole and Wells, 2008). Because vertical momentum is not included, the CE–QUAL–W2 model may give inaccurate results where there is substantial vertical acceleration that may affect the algae groups. The CE–QUAL–W2 model does not have a sediment compartment that simulates kinetics in the sediment and at the sediment-water interface. The simplistic sediment computation in the CE–QUAL–W2 model places a limitation on long-term predictive capabilities of the water-quality part of the CE–QUAL–W2 model. Furthermore, the composition and dynamics of the algal community in a reservoir can be complex. Modeling of the algal dynamics and composition is a large simplification of what actually occurs in a reservoir.

Finally, Lake Maumelle appears to be in its stable trophic equilibrium phase with relatively little variability in trophic response variables (such as nutrients, chlorophyll *a*, and Secchi disk depth). Parameter values determined to describe water-quality processes during 2004–10 may not accurately describe processes as the equilibrium phases end and the reservoir ages. Therefore, the model may not accurately simulate reservoir water quality as watershed conditions become more disturbed.

Simulated Effects of Hydrologic, Water Quality, and Land-Use Changes

The calibrated HSPF and CE–QUAL–W2 models (referred to as "baseline conditions" in the following discussions of scenarios) were used to simulate three land-use scenarios and examine the potential effects of these land-use changes on the water quality of Lake Maumelle if the scenarios were present on the landscape during the 2004 through 2010 simulation period. These scenarios included: scenario 1 that simulated conversion of most land in the watershed to forest, scenario 2 that simulated conversion to low-intensity urban land use in part of the watershed, and scenario 3 that simulated timber harvest in part of the watershed. The scenarios were selected in consultation with CAW personnel and represent a least-disturbed condition (scenario 1) and two potential land-use changes (scenarios 2 and 3) that represent a range of potential generalized land-use changes.

Scenario 1—Simulated Conversion of Most Land in the Lake Maumelle Watershed to Forest

Scenario 1 simulated conversion of most land in the watershed to forest. All areas of anthropogenically altered land, including agriculture area, clearcut area, grassland area, urban land-use area, and all road area, were converted to forest (fig. 29). The converted land use was divided into approximately 70 percent coniferous forest and approximately 30 percent deciduous forest based on the distribution of coniferous to deciduous forest in the baseline model. Converting the urban (0.39 percent), paved roads (0.79 percent), clearcut (5.60 percent), grassland (3.06 percent), and agriculture (0.05 percent) land use (fig. 2) to forest increased the percentage of forested land in the watershed by approximately 9.9 percent (from approximately 79.9 to approximately 89.8 percent).

Simulated land-use changes converting most land in the watershed to forest resulted in little overall effect on the simulated water quality in the HSPF model. Water-quality loading rates simulated by the HSPF model for scenario 1 were generally lower than those for the baseline condition (fig. 30; table 14); the simulated subwatershed loading rates seldom changed more than 10 percent, with the exception of fecal coliform bacteria, with most constituent loading rates decreasing. The largest decreases in water-quality constituents occurred at subwatersheds 25 and 31 (Reece and Yount Creeks, respectively); these subwatersheds had the greatest change from urban to forest land use. Percent differences for the loading rates from the Lake Maumelle watershed, as a whole, ranged from -76.4 percent (fecal coliform bacteria) to -2.9 percent (total organic carbon) (table 14). Percentage differences for individual subwatershed loading rates ranged from -95.7 (fecal coliform bacteria) to 2.6 percent (total phosphorus). As for the entire watershed load, dissolved nitrite plus nitrate and dissolved ammonia decreased 7.2 and 3.6 percent, respectively in scenario 1 relative to the baseline condition, while dissolved orthophosphate and total phosphorous decreased 10.3 and 4.2 percent, respectively (table 14). The Lake Maumelle watershed mean streamflow decreased in response to land-use changes of scenario 1. The simulated mean streamflow into Lake Maumelle decreased 4.4 percent from baseline conditions (table 14) because of the decrease in impervious land, which may explain, in part, the decreases in the nutrient, total organic carbon, and sediment loads.

Base from U.S. Geological Survey digital data, 2008, 1:24,000
Universal Transverse Mercator projection, zone 15

EXPLANATION

Land use and land cover

Water (10.14 percent)

Bare soil (0.06 percent)

Deciduous forest (26.77 percent)

Coniferous forest (63.03 percent)

Watershed boundary

Subwatershed boundary and number

5 — Stream reach (subwatershed number is reach number)

Figure 29. Conversion of most anthropogenically altered land to forest land use for scenario 1.

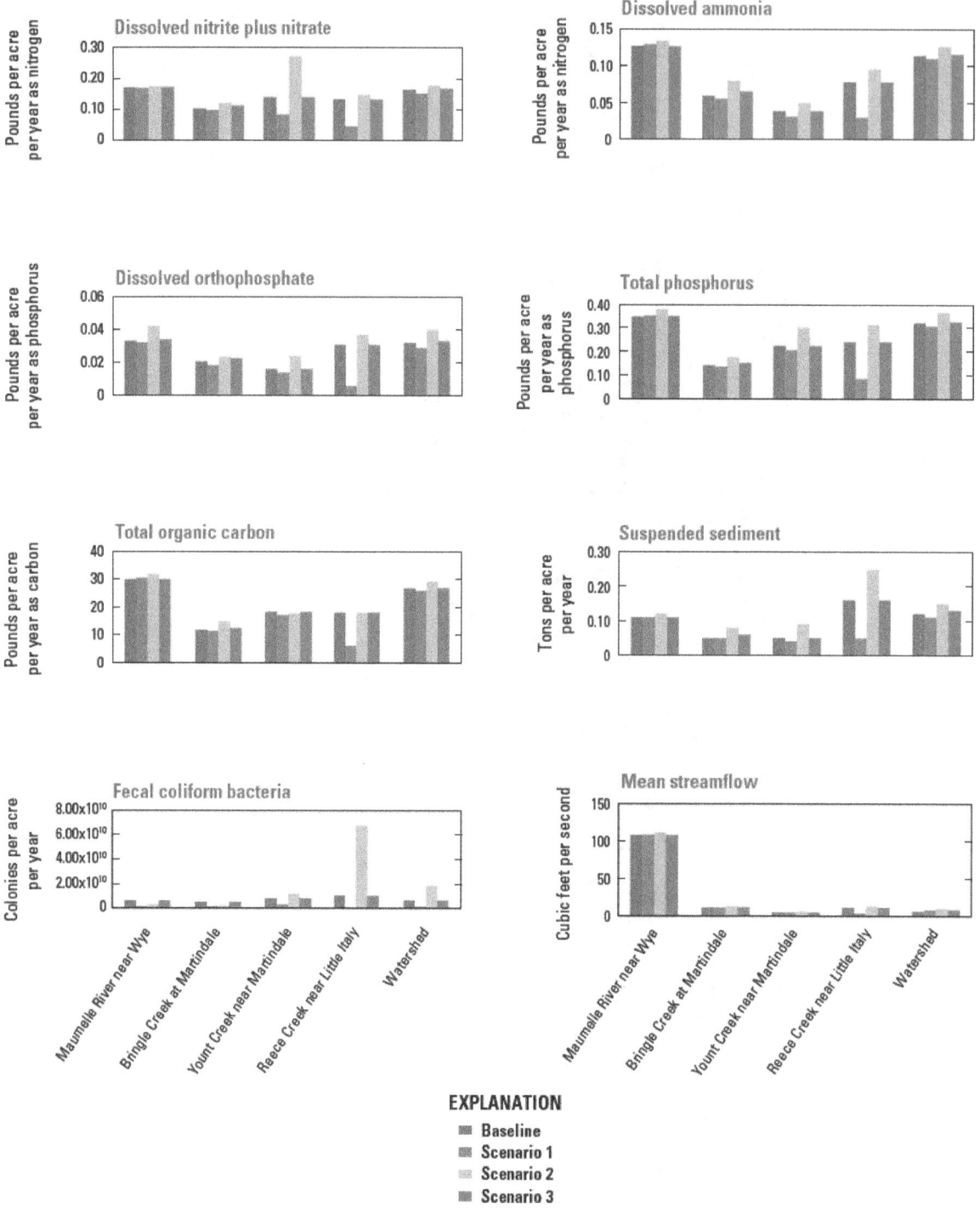

Figure 30.   Histograms showing mean simulated streamflow and median loading rates for baseline condition, scenario 1, scenario 2, and scenario 3 at Maumelle River near Wye, Bringle Creek at Martindale, Yount Creek near Martindale, and Reece Creek near Little Italy.

Table 14. Baseline and scenario loading rates and streamflow comparisons for subwatersheds immediately adjacent to Lake Maumelle from the Hydrological Simulation Program—FORTRAN model of the Lake Maumelle watershed.

[Loading rates for subwatersheds are sums of loads for the listed subwatershed and upstream subwatersheds divided by the acres in these subwatersheds; N, nitrogen; P, phosphorus; C, carbon; (lb/acre)/yr, pound per acre per year; (ton/acre)/yr, ton per acre per year; (col/acre)/yr, colonies per acre per year; ft³/s, cubic feet per second; mean values calculated using years 2004–10; percentage difference equals scenario value minus baseline value divided by baseline value; percentage difference values may be affected by rounding of baseline and scenario values]

| Subwatershed | Mean loading rates | | | | | | | Mean stream-flow (ft³/s) |
	Dissolved nitrite plus nitrate ([lb/acre]/yr as N)	Dissolved ammonia ([lb/acre]/yr as N)	Dissolved orthophos-phate ([lb/acre]/yr as P)	Total phospho-rus ([lb/acre]/yr as P)	Total organic carbon ([lb/acre]/yr as C)	Sus-pended sediment ([ton/acre]/yr)	Fecal coliform bacteria ([col/acre]/yr)	
15 (Maumelle River at Williams Junction; Upper Watershed Area)[1]								
Baseline	0.182	0.138	0.011	0.273	25.7	0.12	3.72×10^9	71.28
Scenario 1	0.181	0.140	0.011	0.280	26.3	0.12	1.55×10^9	71.70
Scenario 2	0.182	0.138	0.011	0.273	25.7	0.12	3.72×10^9	71.28
Scenario 3	0.182	0.137	0.011	0.273	25.6	0.12	3.72×10^9	71.25
Percentage difference between baseline and scenario 1	-0.5	1.4	0.0	2.6	2.3	0.0	-58.2	0.6
Percentage difference between baseline and scenario 2	0.0	0.0	0.0	0.0	0.0	0.0	0.1	0.0
Percentage difference between baseline and scenario 3	0.0	-0.7	0.0	0.0	-0.4	0.0	-0.0	-0.0
18 (Maumelle River near Wye; Critical Area B)[1]								
Baseline	0.171	0.127	0.033	0.346	30.0	0.11	5.96×10^9	108.08
Scenario 1	0.169	0.129	0.032	0.350	30.4	0.11	1.50×10^9	108.47
Scenario 2	0.175	0.134	0.042	0.377	31.8	0.12	3.02×10^9	111.55
Scenario 3	0.172	0.127	0.034	0.348	30	0.11	5.96×10	108.21
Percentage difference between baseline and scenario 1	-1.2	1.6	-3.0	1.2	1.3	0.0	-74.9	0.4
Percentage difference between baseline and scenario 2	2.3	5.5	27.3	9.0	6.0	9.1	-49.3	3.2
Percentage difference between baseline and scenario 3	0.6	0.0	3.0	0.6	0.0	0.0	0.0	0.1
20 (Bringle Creek; Critical Area B)[1]								
Baseline	0.103	0.059	0.020	0.139	11.7	0.05	4.78×10^9	11.38
Scenario 1	0.097	0.055	0.018	0.133	11.3	0.05	1.18×10^9	11.14
Scenario 2	0.120	0.079	0.023	0.172	14.7	0.08	1.57×10^9	12.64
Scenario 3	0.113	0.065	0.022	0.147	12.3	0.06	4.82×10^9	11.71
Percentage difference between baseline and scenario 1	-5.8	-6.8	-10.0	-4.3	-3.4	0.0	-75.4	-2.1
Percentage difference between baseline and scenario 2	16.5	33.9	15.0	23.7	25.6	60.0	-67.2	11.1
Percentage difference between baseline and scenario 3	9.7	10.2	10.0	5.8	5.1	20.0	0.7	2.9
22 (Maumelle River; Critical Area B)								
Baseline	0.178	0.130	0.037	0.355	30.5	0.11	6.18×10^9	123.49
Scenario 1	0.175	0.131	0.036	0.356	30.8	0.11	1.58×10^9	123.55
Scenario 2	0.185	0.141	0.047	0.391	32.8	0.12	3.06×10^9	128.82
Scenario 3	0.180	0.131	0.039	0.358	30.6	0.11	6.18×10	123.98

Table 14.   Baseline and scenario loading rates and streamflow comparisons for subwatersheds immediately adjacent to Lake Maumelle from the Hydrological Simulation Program—FORTRAN model of the Lake Maumelle watershed.—Continued

[Loading rates for subwatersheds are sums of loads for the listed subwatershed and upstream subwatersheds divided by the acres in these subwatersheds; N, nitrogen; P, phosphorus; C, carbon; (lb/acre)/yr, pound per acre per year; (ton/acre)/yr, ton per acre per year; (col/acre)/yr, colonies per acre per year; ft³/s, cubic feet per second; mean values calculated using years 2004–10; percentage difference equals scenario value minus baseline value divided by baseline value; percentage difference values may be affected by rounding of baseline and scenario values]

| Subwatershed | Mean loading rates | | | | | | | Mean stream-flow (ft³/s) |
	Dissolved nitrite plus nitrate ([lb/acre]/yr as N)	Dissolved ammonia ([lb/acre]/yr as N)	Dissolved orthophos-phate ([lb/acre]/yr as P)	Total phospho-rus ([lb/acre]/yr as P)	Total organic carbon ([lb/acre]/yr as C)	Sus-pended sediment ([ton/acre]/yr)	Fecal coliform bacteria ([col/acre]/yr)	
22 (Maumelle River; Critical Area B)—Continued								
Percentage difference between baseline and scenario 1	-1.7	0.8	-2.7	0.3	1.0	0.0	-74.4	0.0
Percentage difference between baseline and scenario 2	3.9	8.5	27.0	10.1	7.5	9.1	-50.4	4.3
Percentage difference between baseline and scenario 3	1.1	0.8	5.4	0.8	0.3	0.0	0.1	0.4
31 (Yount Creek; Critical Area B)								
Baseline	0.140	0.039	0.016	0.225	18.3	0.05	7.89×10^9	4.37
Scenario 1	0.083	0.031	0.014	0.205	17.1	0.04	3.01×10^9	4.26
Scenario 2	0.273	0.050	0.024	0.302	17.8	0.09	1.19×10^{10}	5.60
Scenario 3	0.140	0.039	0.016	0.225	18.3	0.05	7.89×10	4.37
Percentage difference between baseline and scenario 1	-40.7	-20.5	-12.5	-8.9	-6.6	-20.0	-61.8	-2.5
Percentage difference between baseline and scenario 2	95.0	28.2	50.0	34.2	-2.7	80.0	51.2	28.1
Percentage difference between baseline and scenario 3	0.0	0.0	0.0	0.0	0.0	0.0	0.0	0.0
25 (Reece Creek; Critical Area B)								
Baseline	0.133	0.078	0.031	0.240	18.4	0.16	1.04×10^{10}	11.12
Scenario 1	0.046	0.030	0.006	0.085	6.5	0.05	4.48×10^8	3.78
Scenario 2	0.148	0.096	0.037	0.314	18.3	0.25	6.75×10^{10}	13.00
Scenario 3	0.133	0.078	0.031	0.240	18.4	0.16	1.04×10^{10}	11.12
Percentage difference between baseline and scenario 1	-65.4	-61.5	-80.6	-64.6	-64.7	-68.8	-95.7	-66.0
Percentage difference between baseline and scenario 2	11.3	23.1	19.4	30.8	-0.5	56.3	547.8	16.9
Percentage difference between baseline and scenario 3	0.0	0.0	0.0	0.0	0.0	0.0	0.0	0.0
26 (Critical Area B)								
Baseline	0.118	0.090	0.013	0.213	15.9	0.16	1.31×10^9	0.62
Scenario 1	0.118	0.09	0.013	0.213	15.9	0.16	1.26×10^9	0.62
Scenario 2	0.124	0.095	0.014	0.223	16.5	0.18	1.28×10^{10}	0.63
Scenario 3	0.118	0.090	0.013	0.213	15.9	0.16	1.31×10^9	0.62
Percentage difference between baseline and scenario 1	0.0	0.0	0.0	0.0	0.0	0.0	-4.0	0.0
Percentage difference between baseline and scenario 2	5.1	5.6	7.7	4.7	3.8	12.5	877.0	1.6
Percentage difference between baseline and scenario 3	0.0	0.0	0.0	0.0	0.0	0.0	0.0	0.0

Table 14. Baseline and scenario loading rates and streamflow comparisons for subwatersheds immediately adjacent to Lake Maumelle from the Hydrological Simulation Program—FORTRAN model of the Lake Maumelle watershed.—Continued

[Loading rates for subwatersheds are sums of loads for the listed subwatershed and upstream subwatersheds divided by the acres in these subwatersheds; N, nitrogen; P, phosphorus; C, carbon; (lb/acre)/yr, pound per acre per year; (ton/acre)/yr, ton per acre per year; (col/acre)/yr, colonies per acre per year; ft³/s, cubic feet per second; mean values calculated using years 2004–10; percentage difference equals scenario value minus baseline value divided by baseline value; percentage difference values may be affected by rounding of baseline and scenario values]

Subwatershed	Mean loading rates							Mean stream-flow (ft³/s)
	Dissolved nitrite plus nitrate ([lb/acre]/yr as N)	Dissolved ammonia ([lb/acre]/yr as N)	Dissolved orthophos-phate ([lb/acre]/yr as P)	Total phospho-rus ([lb/acre]/yr as P)	Total organic carbon ([lb/acre]/yr as C)	Sus-pended sediment ([ton/acre]/yr)	Fecal coliform bacteria ([col/acre]/yr)	
28 (Critical Area B)								
Baseline	0.138	0.089	0.018	0.259	19.7	0.16	2.87×10^9	3.80
Scenario 1	0.133	0.086	0.017	0.253	19.4	0.15	1.31×10^9	3.77
Scenario 2	0.151	0.105	0.021	0.348	27.0	0.26	7.85×10^{10}	4.59
Scenario 3	0.138	0.089	0.018	0.259	19.7	0.16	2.87×10	3.80
Percentage difference between baseline and scenario 1	-3.6	-3.4	-5.6	-2.3	-1.5	-6.3	-54.3	-0.8
Percentage difference between baseline and scenario 2	9.4	18.0	16.7	34.4	37.1	62.5	2,634.3	20.8
Percentage difference between baseline and scenario 3	0.0	0.0	0.0	0.0	0.0	0.0	0.0	0.0
29 (Critical Area B)								
Baseline	0.081	0.075	0.003	0.189	14.3	0.14	1.01×10^{10}	0.22
Scenario 1	0.074	0.070	0.003	0.180	13.7	0.13	1.16×10^9	0.22
Scenario 2	0.081	0.075	0.003	0.189	14.3	0.14	1.01×10^{10}	0.22
Scenario 3	0.081	0.075	0.003	0.189	14.3	0.14	1.01×10	0.22
Percentage difference between baseline and scenario 1	-8.6	-6.7	0.0	-4.8	-4.2	-7.1	-88.5	0.0
Percentage difference between baseline and scenario 2	0.0	0.0	0.0	0.0	0.0	0.0	0.0	0.0
Percentage difference between baseline and scenario 3	0.0	0.0	0.0	0.0	0.0	0.0	0.0	0.0
33 (Critical Area A)								
Baseline	0.130	0.094	0.014	0.231	17.6	0.16	8.30×10^9	0.91
Scenario 1	0.124	0.088	0.013	0.222	16.9	0.14	1.33×10^9	0.90
Scenario 2	0.136	0.100	0.014	0.246	18.7	0.18	2.30×10^{10}	0.93
Scenario 3	0.130	0.094	0.014	0.231	17.6	0.16	8.30×10^9	0.91
Percentage difference between baseline and scenario 1	-4.6	-6.4	-7.1	-3.9	-4.0	-12.5	-84.0	-1.1
Percentage difference between baseline and scenario 2	4.6	6.4	0.0	6.5	6.2	12.5	177.2	2.2
Percentage difference between baseline and scenario 3	0.0	0.0	0.0	0.0	0.0	0.0	0.0	0.0
35 (Critical Area A)								
Baseline	0.127	0.090	0.010	0.242	18.2	0.17	1.79×10^9	1.32
Scenario 1	0.123	0.086	0.010	0.236	17.9	0.16	1.32×10^9	1.31
Scenario 2	0.140	0.107	0.011	0.295	22.6	0.22	5.06×10	1.46
Scenario 3	0.127	0.090	0.010	0.242	18.2	0.17	1.79×10^9	1.32

Table 14.   Baseline and scenario loading rates and streamflow comparisons for subwatersheds immediately adjacent to Lake Maumelle from the Hydrological Simulation Program—FORTRAN model of the Lake Maumelle watershed.—Continued

[Loading rates for subwatersheds are sums of loads for the listed subwatershed and upstream subwatersheds divided by the acres in these subwatersheds; N, nitrogen; P, phosphorus; C, carbon; (lb/acre)/yr, pound per acre per year; (ton/acre)/yr, ton per acre per year; (col/acre)/yr, colonies per acre per year; ft³/s, cubic feet per second; mean values calculated using years 2004–10; percentage difference equals scenario value minus baseline value divided by baseline value; percentage difference values may be affected by rounding of baseline and scenario values]

| Subwatershed | Mean loading rates | | | | | | | Mean stream-flow (ft³/s) |
	Dissolved nitrite plus nitrate ([lb/acre]/ yr as N)	Dissolved ammonia ([lb/acre]/ yr as N)	Dissolved orthophos-phate ([lb/acre]/ yr as P)	Total phospho-rus ([lb/acre]/ yr as P)	Total organic carbon ([lb/acre]/ yr as C)	Sus-pended sediment ([ton/ acre]/yr)	Fecal coliform bacteria ([col/acre]/yr)	
35 (Critical Area A)—Continued								
Percentage difference between baseline and scenario 1	-3.1	-4.4	0.0	-2.5	-1.6	-5.9	-26.4	-0.8
Percentage difference between baseline and scenario 2	10.2	18.9	10.0	21.9	24.2	29.4	2,726.2	10.6
Percentage difference between baseline and scenario 3	0.0	0.0	0.0	0.0	0.0	0.0	0.0	0.0
37 (Critical Area B)								
Baseline	0.135	0.088	0.018	0.247	18.7	0.16	5.70×10^9	4.42
Scenario 1	0.134	0.087	0.018	0.245	18.6	0.16	1.27×10^9	4.41
Scenario 2	0.153	0.113	0.021	0.328	25.2	0.25	6.84×10	5.21
Scenario 3	0.135	0.088	0.018	0.247	18.7	0.16	5.70×10^9	4.42
Percentage difference between baseline and scenario 1	-0.7	-1.1	0.0	-0.8	-0.5	0.0	-77.6	-0.2
Percentage difference between baseline and scenario 2	13.3	28.4	16.7	32.8	34.8	56.3	1,100.6	17.9
Percentage difference between baseline and scenario 3	0.0	0.0	0.0	0.0	0.0	0.0	0.0	0.0
39 (Critical Area A)								
Baseline	0.139	0.102	0.012	0.248	18.9	0.16	3.79×10^9	1.26
Scenario 1	0.138	0.100	0.011	0.246	18.8	0.16	1.41×10^9	1.26
Scenario 2	0.140	0.102	0.012	0.248	18.9	0.16	4.30×10^9	1.26
Scenario 3	0.139	0.102	0.012	0.248	18.9	0.16	3.79×10^9	1.26
Percentage difference between baseline and scenario 1	-0.7	-2.0	-8.3	-0.8	-0.5	0.0	-62.9	0.0
Percentage difference between baseline and scenario 2	0.7	0.0	0.0	0.0	0.0	0.0	13.3	0.0
Percentage difference between baseline and scenario 3	0.0	0.0	0.0	0.0	0.0	0.0	0.0	0.0
41 (Critical Area A)								
Baseline	0.139	0.098	0.012	0.248	18.8	0.17	1.66×10^9	1.63
Scenario 1	0.139	0.098	0.012	0.248	18.7	0.17	1.38×10^9	1.63
Scenario 2	0.139	0.098	0.012	0.248	18.8	0.17	1.81×10^9	1.63
Scenario 3	0.139	0.098	0.012	0.248	18.8	0.17	1.66×10^9	1.63
Percentage difference between baseline and scenario 1	0.0	0.0	0.0	0.0	-0.5	0.0	-16.7	0.0
Percentage difference between baseline and scenario 2	0.0	0.0	0.0	0.0	0.0	0.0	9.3	0.0
Percentage difference between baseline and scenario 3	0.0	0.0	0.0	0.0	0.0	0.0	0.0	0.0

Table 14. Baseline and scenario loading rates and streamflow comparisons for subwatersheds immediately adjacent to Lake Maumelle from the Hydrological Simulation Program—FORTRAN model of the Lake Maumelle watershed.—Continued

[Loading rates for subwatersheds are sums of loads for the listed subwatershed and upstream subwatersheds divided by the acres in these subwatersheds; N, nitrogen; P, phosphorus; C, carbon; (lb/acre)/yr, pound per acre per year; (ton/acre)/yr, ton per acre per year; (col/acre)/yr, colonies per acre per year; ft³/s, cubic feet per second; mean values calculated using years 2004–10; percentage difference equals scenario value minus baseline value divided by baseline value; percentage difference values may be affected by rounding of baseline and scenario values]

	Mean loading rates							Mean stream-flow (ft³/s)
Subwatershed	Dissolved nitrite plus nitrate ([lb/acre]/ yr as N)	Dissolved ammonia ([lb/acre]/ yr as N)	Dissolved orthophos-phate ([lb/acre]/ yr as P)	Total phospho-rus ([lb/acre]/ yr as P)	Total organic carbon ([lb/acre]/ yr as C)	Sus-pended sediment ([ton/ acre]/yr)	Fecal coliform bacteria ([col/acre]/yr)	
43 (Critical Area A)								
Baseline	0.141	0.095	0.018	0.254	19.3	0.16	1.60×10¹⁰	2.76
Scenario 1	0.140	0.094	0.017	0.253	19.3	0.16	1.32×10⁹	2.76
Scenario 2	0.141	0.095	0.018	0.255	19.4	0.17	1.71×10¹⁰	2.76
Scenario 3	0.141	0.095	0.018	0.254	19.3	0.16	1.60×10¹⁰	2.76
Percentage difference between baseline and scenario 1	-0.7	-1.1	-5.6	-0.4	0.0	0.0	-91.8	0.0
Percentage difference between baseline and scenario 2	0.0	0.0	0.0	0.4	0.5	6.3	6.8	0.0
Percentage difference between baseline and scenario 3	0.0	0.0	0.0	0.0	0.0	0.0	0.0	0.0
45 (Critical Area B)								
Baseline	0.140	0.095	0.023	0.248	19.4	0.13	7.60×10⁹	2.28
Scenario 1	0.139	0.094	0.023	0.247	19.4	0.13	1.34×10⁹	2.27
Scenario 2	0.149	0.105	0.025	0.273	21.2	0.17	2.47×10¹⁰	2.36
Scenario 3	0.143	0.097	0.024	0.253	19.7	0.14	7.61×10⁹	2.29
Percentage difference between baseline and scenario 1	-0.7	-1.1	0.0	-0.4	0.0	0.0	-82.3	-0.4
Percentage difference between baseline and scenario 2	6.4	10.5	8.7	10.1	9.3	30.8	225.3	3.5
Percentage difference between baseline and scenario 3	2.1	2.1	4.3	2.0	1.5	7.7	0.2	0.4
47 (Critical Area A)								
Baseline	0.136	0.097	0.017	0.246	18.8	0.16	1.06×10¹⁰	0.89
Scenario 1	0.132	0.094	0.016	0.240	18.3	0.15	1.42×10⁹	0.89
Scenario 2	0.138	0.099	0.017	0.248	18.9	0.17	1.34×10¹⁰	0.90
Scenario 3	0.136	0.097	0.017	0.246	18.8	0.16	1.06×10¹⁰	0.89
Percentage difference between baseline and scenario 1	-2.9	-3.1	-5.9	-2.4	-2.7	-6.3	-86.7	0.0
Percentage difference between baseline and scenario 2	1.5	2.1	0.0	0.8	0.5	6.3	26.0	1.1
Percentage difference between baseline and scenario 3	0.0	0.0	0.0	0.0	0.0	0.0	0.0	0.0
49 (Critical Area B)								
Baseline	0.141	0.086	0.025	0.267	20.6	0.17	2.37×10⁹	4.48
Scenario 1	0.135	0.082	0.024	0.259	20.2	0.15	1.37×10⁹	4.44
Scenario 2	0.157	0.111	0.029	0.325	25.3	0.24	5.57×10¹⁰	4.96
Scenario 3	0.152	0.094	0.028	0.281	21.2	0.20	2.40×10⁹	4.55

Table 14.   Baseline and scenario loading rates and streamflow comparisons for subwatersheds immediately adjacent to Lake Maumelle from the Hydrological Simulation Program—FORTRAN model of the Lake Maumelle watershed.—Continued

[Loading rates for subwatersheds are sums of loads for the listed subwatershed and upstream subwatersheds divided by the acres in these subwatersheds; N, nitrogen; P, phosphorus; C, carbon; (lb/acre)/yr, pound per acre per year; (ton/acre)/yr, ton per acre per year; (col/acre)/yr, colonies per acre per year; ft³/s, cubic feet per second; mean values calculated using years 2004–10; percentage difference equals scenario value minus baseline value divided by baseline value; percentage difference values may be affected by rounding of baseline and scenario values]

| | Mean loading rates | | | | | | | Mean stream-flow (ft³/s) |
Subwatershed	Dissolved nitrite plus nitrate ([lb/acre]/ yr as N)	Dissolved ammonia ([lb/acre]/ yr as N)	Dissolved orthophos-phate ([lb/acre]/ yr as P)	Total phospho-rus ([lb/acre]/ yr as P)	Total organic carbon ([lb/acre]/ yr as C)	Sus-pended sediment ([ton/ acre]/yr)	Fecal coliform bacteria ([col/acre]/yr)	
49 (Critical Area B)—Continued								
Percentage difference between baseline and scenario 1	-4.3	-4.7	-4.0	-3.0	-1.9	-11.8	-42.2	-0.9
Percentage difference between baseline and scenario 2	11.3	29.1	16.0	21.7	22.8	41.2	2,252.7	10.7
Percentage difference between baseline and scenario 3	7.8	9.3	12.0	5.2	2.9	17.6	1.2	1.6
51 (Critical Area B)								
Baseline	0.136	0.088	0.019	0.255	19.6	0.15	4.11×10^9	3.78
Scenario 1	0.134	0.086	0.019	0.253	19.5	0.15	1.34×10^9	3.78
Scenario 2	0.152	0.110	0.022	0.331	25.5	0.25	6.35×10^{10}	4.38
Scenario 3	0.162	0.102	0.022	0.289	21.3	0.22	4.18×10^9	3.94
Percentage difference between baseline and scenario 1	-1.5	-2.3	0.0	-0.8	-0.5	0.0	-67.3	0.0
Percentage difference between baseline and scenario 2	11.8	25.0	15.8	29.8	30.1	66.7	1,446.8	15.9
Percentage difference between baseline and scenario 3	19.1	15.9	15.8	13.3	8.7	46.7	1.8	4.2
53 (Critical Area B)								
Baseline	0.141	0.090	0.019	0.273	20.9	0.17	3.46×10^9	4.24
Scenario 1	0.132	0.082	0.018	0.261	20.2	0.15	1.34×10^9	4.19
Scenario 2	0.159	0.119	0.022	0.374	28.9	0.29	8.98×10^{10}	5.10
Scenario 3	0.160	0.105	0.021	0.297	22.1	0.22	3.51×10^9	4.35
Percentage difference between baseline and scenario 1	-6.4	-8.9	-5.3	-4.4	-3.3	-11.8	-61.2	-1.2
Percentage difference between baseline and scenario 2	12.8	32.2	15.8	37.0	38.3	70.6	2,493.8	20.3
Percentage difference between baseline and scenario 3	13.5	16.7	10.5	8.8	5.7	29.4	1.5	2.6
Watershed								
Loading rate and streamflow for entire watershed, baseline[2,3]	0.164	0.114	0.032	0.320	26.7	0.12	6.34×10^9	8.58
Loading rate and streamflow for entire watershed scenario 1[2,3]	0.153	0.110	0.029	0.307	26.0	0.11	1.50×10^9	8.20
Loading rate and streamflow for entire watershed scenario 2[2,3]	0.178	0.127	0.040	0.364	29.2	0.15	1.86×10^{10}	9.19
Loading rate and streamflow for entire watershed scenario 3[2,3]	0.168	0.116	0.033	0.324	26.9	0.13	6.35×10	8.62

Table 14. Baseline and scenario loading rates and streamflow comparisons for subwatersheds immediately adjacent to Lake Maumelle from the Hydrological Simulation Program—FORTRAN model of the Lake Maumelle watershed.—Continued

[Loading rates for subwatersheds are sums of loads for the listed subwatershed and upstream subwatersheds divided by the acres in these subwatersheds; N, nitrogen; P, phosphorus; C, carbon; (lb/acre)/yr, pound per acre per year; (ton/acre)/yr, ton per acre per year; (col/acre)/yr, colonies per acre per year; ft³/s, cubic feet per second; mean values calculated using years 2004–10; percentage difference equals scenario value minus baseline value divided by baseline value; percentage difference values may be affected by rounding of baseline and scenario values]

Subwatershed	Mean loading rates							Mean stream-flow (ft³/s)
	Dissolved nitrite plus nitrate ([lb/acre]/yr as N)	Dissolved ammonia ([lb/acre]/yr as N)	Dissolved orthophos-phate ([lb/acre]/yr as P)	Total phospho-rus ([lb/acre]/yr as P)	Total organic carbon ([lb/acre]/yr as C)	Sus-pended sediment ([ton/acre]/yr)	Fecal coliform bacteria ([col/acre]/yr)	
Watershed—Continued								
Percentage difference between baseline and scenario 1	-7.2	-3.6	-10.3	-4.2	-2.9	-9.8	-76.4	-4.4
Percentage difference between baseline and scenario 2	8.5	11.4	25.0	13.8	9.3	27.3	193.8	7.1
Percentage difference between baseline and scenario 3	2.4	1.8	3.1	1.3	0.7	4.7	0.1	0.5
Scenario 2 (affected subwatersheds)								
Baseline loading rate and streamflow, subwatersheds[4]: 22,25,26,28,31,33,35,37,39,43, 45,47,49,51	0.165	0.115	0.033	0.322	27.0	0.12	6.47×10⁹	11.82
Scenario 2 loading rate and streamflow, subwatersheds[4]: 22,25,26,28,31,33,35,37,39,43, 45,47,49,51	0.179	0.128	0.041	0.366	29.4	0.15	1.68×10¹⁰	12.63
Percentage difference between baseline and affected subwatersheds for scenario 2	8.1	11.0	24.4	13.6	8.8	25.7	160.3	6.9
Scenario 3 (affected subwatersheds)								
Baseline loading rate and streamflow, subwatersheds[4]: 22,45,49,51,53	0.147	0.098	0.025	0.280	22.2	0.15	4.74×10⁹	27.66
Scenario 3 loading rate and streamflow, subwatersheds[4]: 22,45,49,51,53	0.159	0.106	0.027	0.296	23.0	0.18	4.78×10⁹	27.82
Percentage difference between baseline and affected subwatersheds for scenario 3	8.3	8.2	8.9	5.7	3.5	21.9	0.7	0.6

[1]Components (upstream subwatersheds) of subwatershed 22.

[2]To summarize the loading rate for the entire watershed, the constituent loads for all subwatersheds were summed together, then divided by the total acres present within the entire watershed.

[3]Watershed does not include Lake Maumelle surface.

[4]To summarize the loading rate for the group of affected subwatersheds, the constituent loads for affected subwatersheds were summed together, then divided by the total acres present within the group of affected subwatersheds.

The simulated changes in nutrient, suspended sediment, total organic carbon, and fecal coliform bacteria loads from the HSPF model from scenario 1 (fig. 30; table 14) resulted in very slight changes in simulated water quality for Lake Maumelle, relative to the baseline condition for both the epilimnion and hypolimnion (fig. 31a–b). Nutrients, both nitrogen and phosphorus concentrations, on average, over the simulation period (2004–10) increased slightly (fig. 31a; table 15). This resulted in a slight increase in chlorophyll a concentrations at all three lake stations. However, these increases were an order of magnitude (or greater) lower than the RMSE and MAE between measured and predicted concentrations in the baseline model. In late October 2006, when the pool altitude was at its lowest during the simulation period (2004–10), about 7 ft below spillway (October 19, 2006), following complete lake mixing, phosphorus and nitrogen concentrations were at their highest; chlorophyll a responded accordingly at Natural Steps (fig. 32). Similar results were found at the other two stations. The increases in nutrient and chlorophyll a concentrations during late October 2006 and into 2007 were enough to increase these concentrations, on average, over the simulation period (2004–10; fig. 31a). In the remainder of the simulation period (late 2007–10), nutrients and chlorophyll a concentrations in Scenario 1 were similar to the baseline condition.

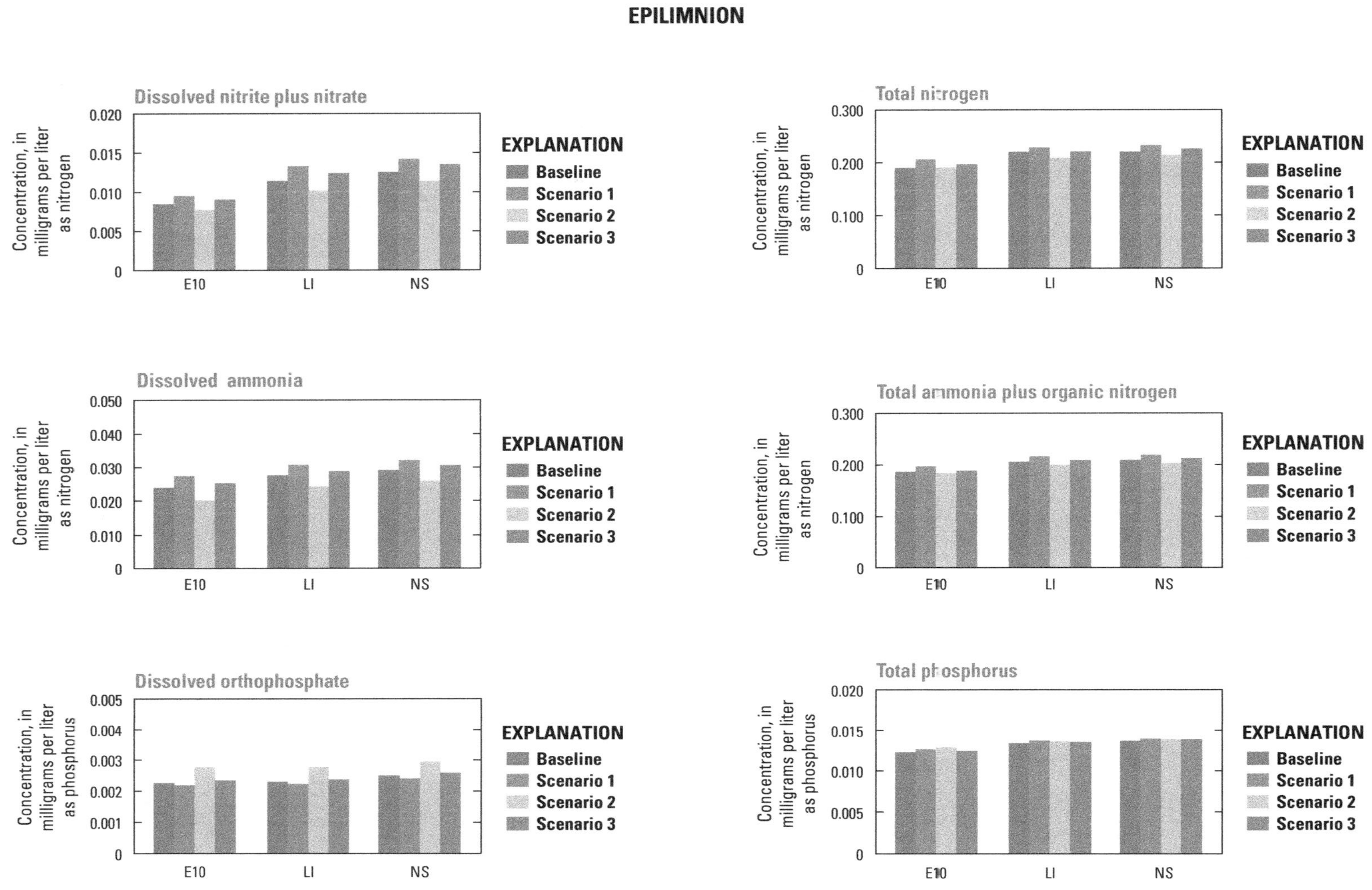

EPILIMNION

Figure 31a. Mean simulated constituent concentrations and Secchi disk depth for epilimnion baseline condition, scenario 1, scenario 2, and scenario 3 at Lake Maumelle east of Highway 10 bridge (E10), Lake Maumelle near Little Italy (LI), and Lake Maumelle near Natural Steps (NS).

EPILIMNION

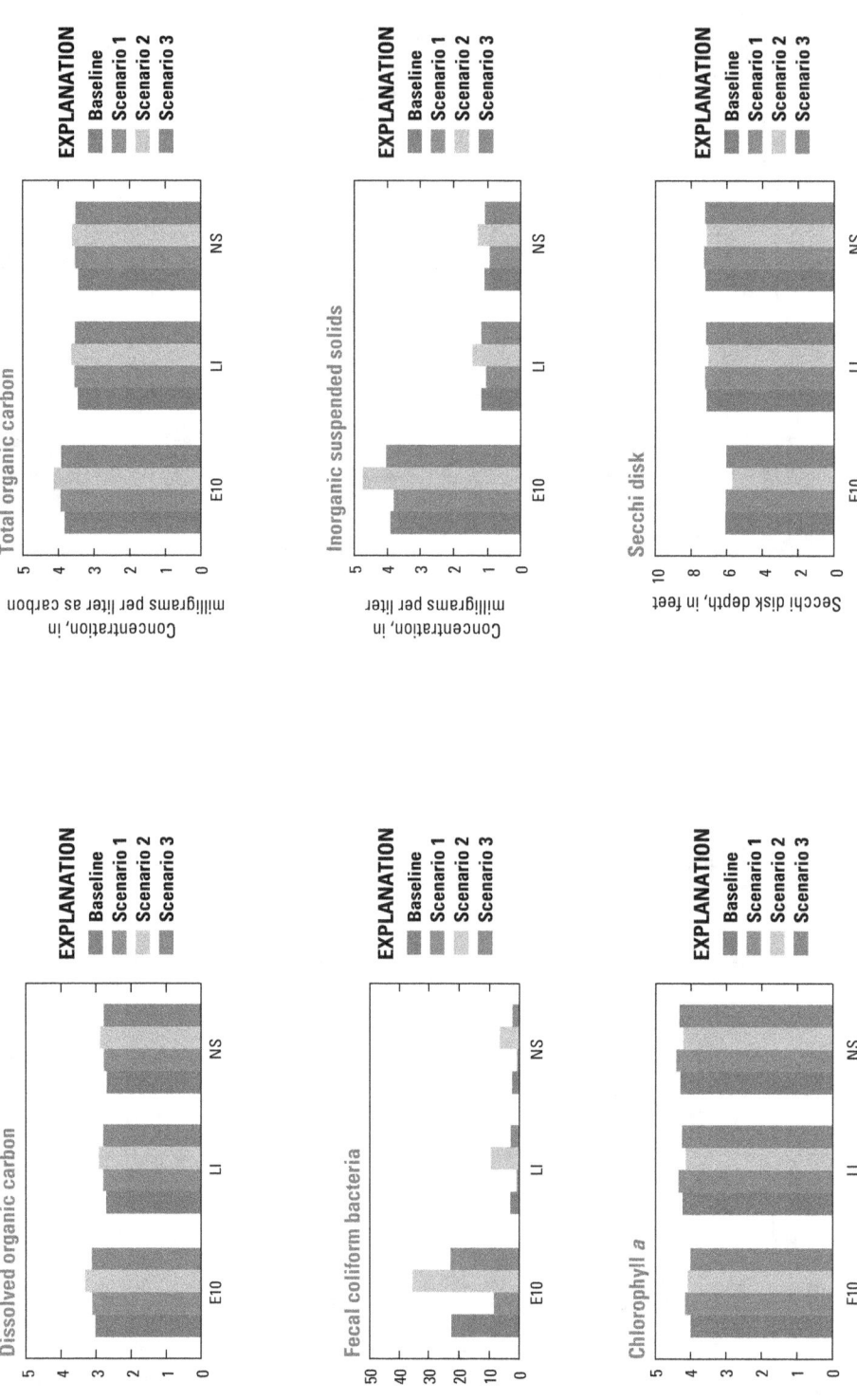

Figure 31a. Mean simulated constituent concentrations and Secchi disk depth for epilimnion baseline condition, scenario 1, scenario 2, and scenario 3 at Lake Maumelle east of Highway 10 bridge (E10), Lake Maumelle near Little Italy (LI), and Lake Maumelle near Natural Steps (NS).—Continued

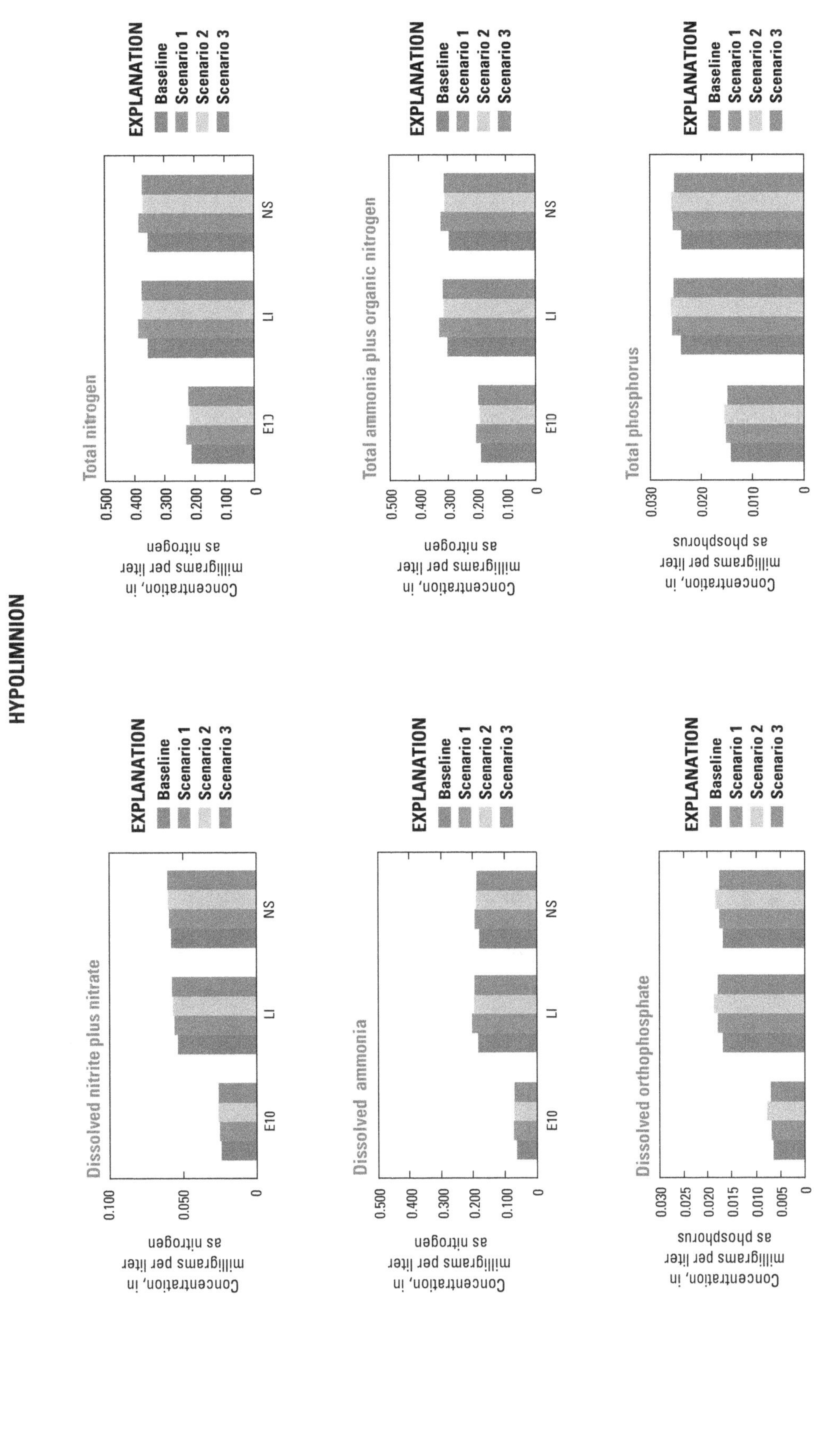

HYPOLIMNION

Total nitrogen

Concentration, in milligrams per liter as nitrogen

EXPLANATION
Baseline
Scenario 1
Scenario 2
Scenario 3

E10   LI   NS

Total ammonia plus organic nitrogen

Concentration, in milligrams per liter as nitrogen

EXPLANATION
Baseline
Scenario 1
Scenario 2
Scenario 3

E10   LI   NS

Total phosphorus

Concentration, in milligrams per liter as phosphorus

EXPLANATION
Baseline
Scenario 1
Scenario 2
Scenario 3

E10   LI   NS

Dissolved nitrite plus nitrate

Concentration, in milligrams per liter as nitrogen

EXPLANATION
Baseline
Scenario 1
Scenario 2
Scenario 3

E10   LI   NS

Dissolved ammonia

Concentration, in milligrams per liter as nitrogen

EXPLANATION
Baseline
Scenario 1
Scenario 2
Scenario 3

E10   LI   NS

Dissolved orthophosphate

Concentration, in milligrams per liter as phosphorus

EXPLANATION
Baseline
Scenario 1
Scenario 2
Scenario 3

E10   LI   NS

Figure 31b. Mean simulated constituent concentrations and Secchi disk depth for hypolimnion baseline condition, scenario 1, scenario 2, and scenario 3 at Lake Maumelle east of Highway 10 bridge (E10), Lake Maumelle near Little Italy (LI), and Lake Maumelle near Natural Steps (NS).

HYPOLIMNION

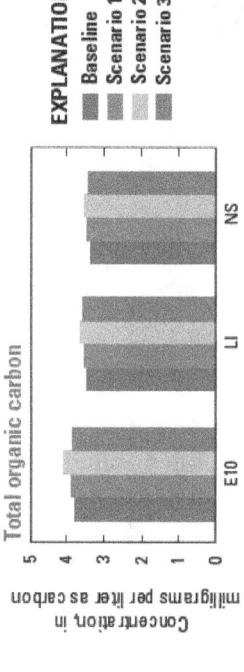

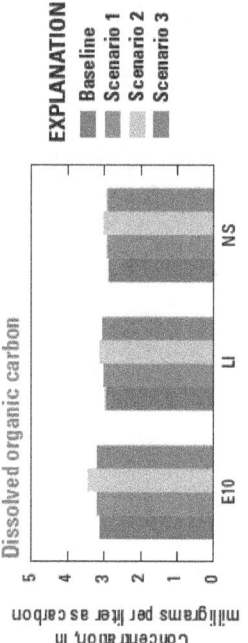

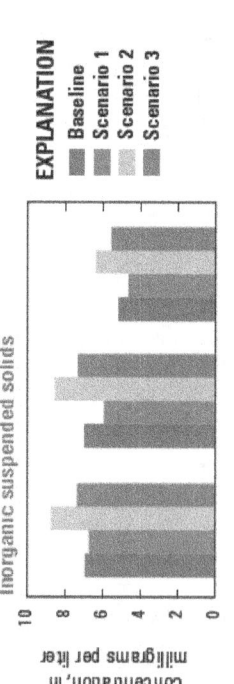

Figure 31b.   Mean simulated constituent concentrations and Secchi disk depth for hypolimnion baseline condition, scenario 1, scenario 2, and scenario 3 at Lake Maumelle east of Highway 10 bridge (E10), Lake Maumelle near Little Italy (LI), and Lake Maumelle near Natural Steps (NS).—Continued

Table 15. Mean simulated water-quality concentrations for baseline condition, scenario 1, scenario 2, and scenario 3 at Lake Maumelle east of Highway 10 bridge, Lake Maumelle near Little Italy, and Lake Maumelle near Natural Steps.

[N, nitrogen; P, phosphorus; C, carbon; mg/L, milligrams per liter; µg/L, micrograms per liter; col/100 mL, colonies per 100 milliliters; ft, feet; LRL, laboratory reporting limit]

CE–QUAL–W2 simulated baseline

Lake station	Layer	Dissolved nitrite plus nitrate (mg/L as N)[1]	Total nitrogen (mg/L as N)[2]	Dissolved ammonia (mg/L as N)[3]	Total ammonia plus organic nitrogen (mg/L as N)[4]	Dissolved orthophosphate (mg/L as P)[5]	Total phosphorus (mg/L as P)[6]	Dissolved organic carbon (mg/L as C)[7]	Total organic carbon (mg/L as C)[8]	Fecal coliform (col/100mL)[9]	Inorganic suspended solids (mg/L)[10]	Chlorophyll a[11] (µg/L)	Secchi disk depth (ft)[12]
East of Highway 10	Epilimnion	0.008	0.19	0.02	0.19	0.002	0.012	3.0	3.8	22	3.90	4.0	6.0
	Hypolimnion	0.024	0.21	0.07	0.19	0.006	0.014	3.1	3.7	40	6.96		
Little Italy	Epilimnion	0.011	0.22	0.03	0.21	0.002	0.013	2.7	3.4	3	1.18	4.2	7.1
	Hypolimnion	0.053	0.36	0.18	0.30	0.017	0.024	2.9	3.5	27	7.02		
Natural Steps	Epilimnion	0.013	0.22	0.03	0.21	0.002	0.014	2.7	3.4	2	1.08	4.3	7.2
	Hypolimnion	0.058	0.36	0.18	0.30	0.017	0.024	2.9	3.4	18	5.19		

CE–QUAL–W2 simulated scenario 1

Lake station	Layer	Dissolved nitrite plus nitrate (mg/L as N)[1]	Total nitrogen (mg/L as N)[2]	Dissolved ammonia (mg/L as N)[3]	Total ammonia plus organic nitrogen (mg/L as N)[4]	Dissolved orthophosphate (mg/L as P)[5]	Total phosphorus (mg/L as P)[6]	Dissolved organic carbon (mg/L as C)[7]	Total organic carbon (mg/L as C)[8]	Fecal coliform (col/100mL)[9]	Inorganic suspended solids (mg/L)[10]	Chlorophyll a[11] (µg/L)	Secchi disk depth (ft)[12]
East of Highway 10	Epilimnion	0.010	0.21	0.03	0.20	0.002	0.013	3.1	3.9	8	3.81	4.2	6.0
	Hypolimnion	0.025	0.23	0.07	0.20	0.007	0.015	3.2	3.9	16	6.74		
Little Italy	Epilimnion	0.013	0.23	0.03	0.22	0.002	0.014	2.8	3.5	1	1.04	4.3	7.2
	Hypolimnion	0.056	0.39	0.20	0.33	0.018	0.026	3.0	3.5	5	5.95		
Natural Steps	Epilimnion	0.014	0.23	0.03	0.22	0.002	0.014	2.8	3.5	1	0.94	4.4	7.3
	Hypolimnion	0.059	0.39	0.19	0.33	0.017	0.026	2.9	3.4	3	4.66		

Table 15. Mean simulated water-quality concentrations for baseline condition, scenario 1, scenario 2, and scenario 3 at Lake Maumelle east of Highway 10 bridge, Lake Maumelle near Little Italy, and Lake Maumelle near Natural Steps.—Continued

[N, nitrogen; P, phosphorus; C, carbon; mg/L, milligrams per liter; µg/L, micrograms per liter; col/100 mL, colonies per 100 milliliters; ft, feet; LRL, laboratory reporting limit]

CE–QUAL–W2 simulated scenario 2

Lake station	Layer	Dissolved nitrite plus nitrate (mg/L as N)[1]	Total nitrogen (mg/L as N)[2]	Dissolved ammonia (mg/L as N)[3]	Total ammonia plus organic nitrogen (mg/L as N)[4]	Dissolved orthophosphate (mg/L as P)[5]	Total phosphorus (mg/L as P)[6]	Dissolved organic carbon (mg/L as C)[7]	Total organic carbon (mg/L as C)[8]	Fecal coliform (col/100mL)[9]	Inorganic suspended solids (mg/L)[10]	Chlorophyll a[11] (µg/L)	Secchi disk depth (ft)[12]
East of Highway 10	Epilimnion	0.008	0.19	0.02	0.18	0.003	0.013	3.3	4.1	36	4.75	4.1	5.7
	Hypolimnion	0.026	0.22	0.07	0.19	0.008	0.015	3.4	4.1	130	8.78		
Little Italy	Epilimnion	0.010	0.21	0.02	0.20	0.003	0.014	2.9	3.6	9	1.44	4.1	7.0
	Hypolimnion	0.057	0.37	0.20	0.32	0.019	0.026	3.1	3.6	92	8.59		
Natural Steps	Epilimnion	0.011	0.21	0.03	0.20	0.003	0.014	2.9	3.6	7	1.29	4.2	7.1
	Hypolimnion	0.060	0.37	0.19	0.31	0.018	0.026	3.0	3.5	49	6.41		

CE–QUAL–W2 simulated scenario 3

Lake station	Layer	Dissolved nitrite plus nitrate (mg/L as N)[1]	Total nitrogen (mg/L as N)[2]	Dissolved ammonia (mg/L as N)[3]	Total ammonia plus organic nitrogen (mg/L as N)[4]	Dissolved orthophosphate (mg/L as P)[5]	Total phosphorus (mg/L as P)[6]	Dissolved organic carbon (mg/L as C)[7]	Total organic carbon (mg/L as C)[8]	Fecal coliform (col/100mL)[9]	Inorganic suspended solids (mg/L)[10]	Chlorophyll a[11] (µg/L)	Secchi disk depth (ft)[12]
East of Highway 10	Epilimnion	0.009	0.20	0.03	0.19	0.002	0.012	3.1	3.9	23	4.04	4.0	6.0
	Hypolimnion	0.026	0.22	0.07	0.20	0.007	0.015	3.2	3.8	40	7.39		
Little Italy	Epilimnion	0.012	0.22	0.03	0.21	0.002	0.014	2.8	3.5	3	1.17	4.3	7.1
	Hypolimnion	0.057	0.38	0.20	0.32	0.018	0.025	3.0	3.5	27	7.35		
Natural Steps	Epilimnion	0.014	0.23	0.03	0.21	0.003	0.014	2.8	3.5	2	1.07	4.3	7.2
	Hypolimnion	0.060	0.37	0.19	0.31	0.017	0.025	2.9	3.4	16	5.58		

[1]Dissolved nitrite plus nitrate LRL: 0.016 mg/L; after October 1, 2010, LRL: 0.008 mg/L.

[2]Total nitrogen LRL: not applicable.

[3]Dissolved ammonia LRL: 0.01 mg/L; after October 1, 2006, LRL: 0.02 mg/L.

[4]Total ammonia plus organic nitrogen LRL: 0.1 mg/L.

[5]Dissolved orthophosphate LRL: 0.006 mg/L; after October 1, 2008, LRL: 0.008 mg/L.

[6]Total phosphorus LRL: 0.004 mg/L; after October 1, 2008, LRL: 0.008 mg/L.

[7]Dissolved organic carbon LRL: 0.3 mg/L.

[8]Total organic carbon LRL: 0.40 mg/L; after October 1, 2008, LRL: 0.60 mg/L.

[9]Fecal coliform LRL: not applicable.

[10]Inorganic suspended solids LRL: 1 mg/L.

[11]Chlorophyll a LRL: 0.1 (µg/L).

[12]Secchi disc LRL: not applicable.

Segment 19,
NATURAL STEPS

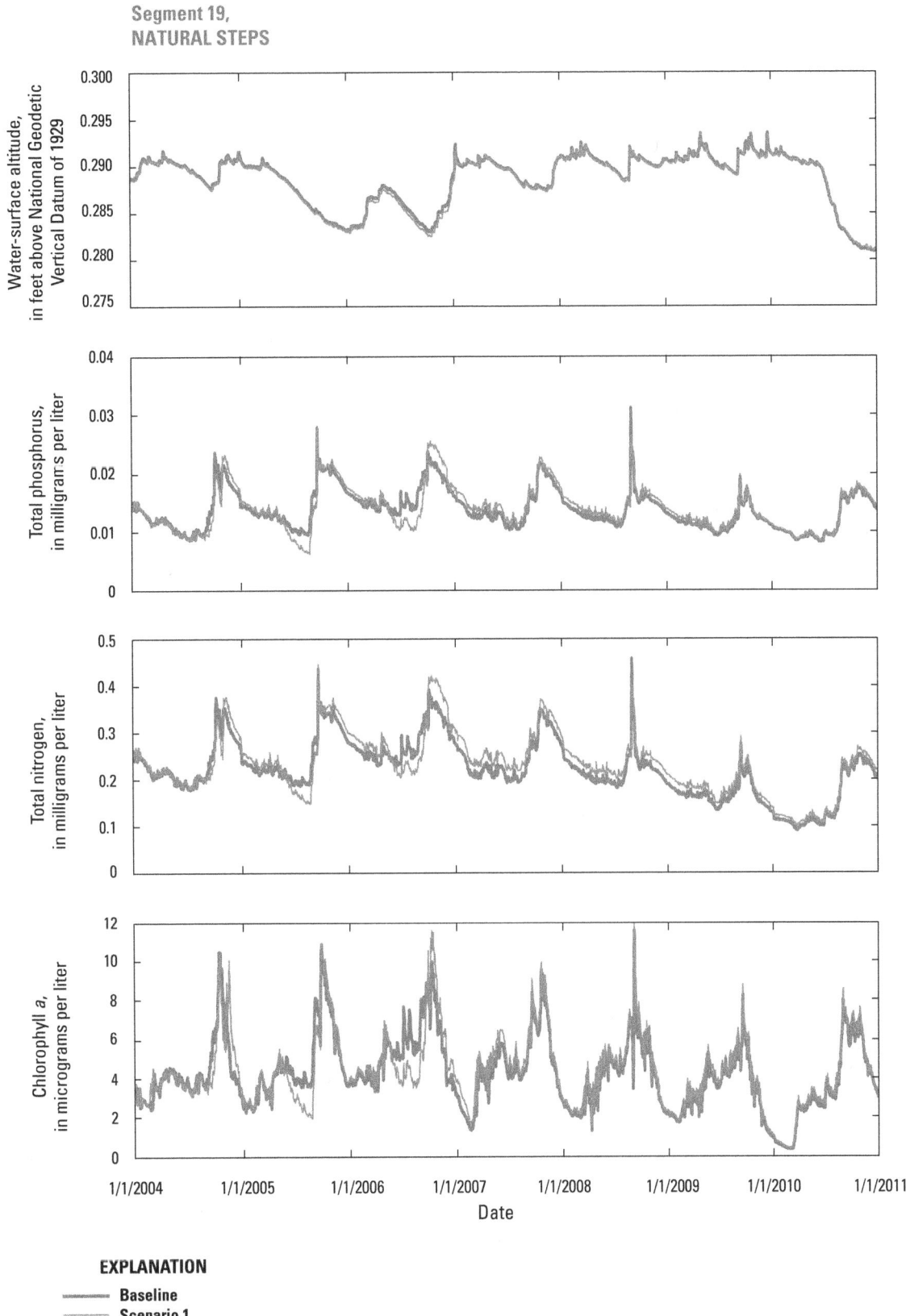

EXPLANATION
— Baseline
— Scenario 1

Figure 32.   Time-series plots of water-surface altitude, total phosphorus, total nitrogen, and chlorophyll *a* concentrations from scenario 1 results compared against the baseline condition.

Scenario 2—Simulated Conversion of Selected Developable Land to Low-Intensity Urban

Scenario 2 included converting all of the land that is considered developable (land in private ownership with a slope of less than 25 percent; data from Vince Guillet, Central Arkansas Water, written commun., 2009) in Pulaski County to low-intensity urban land use. Converting all clearcut, bare soil, grasslands, agriculture, and forest land use (which is not owned by a government entity) into low-intensity urban (fig. 33), using calibrated baseline low-intensity urban land-use parameters, increased the percentage of low-intensity urban land by approximately 20 percent (approximately 17,500 acres) of the total acreage (86,907 acres) for the watershed. The low-intensity urban land-use scenario is not meant to represent any particular housing density. It is simply an increase of existing (baseline) low-intensity urban landscapes (housing densities, lot size, and impervious area) to approximately 20 percent of the watershed, reducing forest and other landscapes by the same.

To accommodate for nutrients in septic systems within the watershed, higher concentrations of nitrogen, ammonia, and phosphorus in the interflow and groundwater were applied to the urban land use in all subwatersheds for the baseline model. This original higher concentration allowed for an increase in the same nutrients for scenario 2. Dissolved nitrite plus nitrate nitrogen concentrations that were associated with interflow and groundwater for deciduous and coniferous forest land use were increased by 30 percent and 45 percent, respectively, for urban land use. These percent increases were determined from studies indicating that approximately 36 percent of total nitrogen loads are derived from septic system leachate (Byron and Burchard, 2008; Maryland Department of the Environment, 2011). Fecal coliform bacteria were handled in a similar fashion as nutrients for scenario 2. Additionally, it was assumed that new septic systems would be installed with new development, and the chance of failure of these systems would be slight. However, within the baseline model, the ACCUM and SQOLIM (table 9) parameter values for fecal coliform bacteria were larger for urban land than for nonurban. These larger parameter values allowed for an increase, and in some instances a very large increase, in the number of colonies per acre per year for affected subwatersheds (table 14) and accounted for septic systems already in place within the watershed.

The land-use change of scenario 2 affected parts of the Bringle, Reece, and Yount Creek subwatersheds, as well as most other subwatersheds that drain into the northern side of Lake Maumelle and the downstream part of the Maumelle River watershed. Water quality and flows at Williams Junction were not affected by the land-use change in scenario 2 because of the location of Williams Junction within the watershed. All of the reaches that drain the subwatersheds above the Williams Junction station lie outside Pulaski County and, therefore, the land use within these subwatersheds was not affected by the conversion to low-intensity urban.

Simulated water-quality loading rates increased in the subwatersheds that include Bringle, Reece, and Yount Creeks in scenario 2 relative to the baseline condition (fig. 30; table 14). These subwatersheds and most others that drain into the northern side of Lake Maumelle had increases greater than 10 percent for nitrite plus nitrate nitrogen, dissolved ammonia nitrogen, dissolved orthophosphate, total phosphorus, and suspended sediment loading rates. The largest overall percentage increase, approximately 2,726 percent, for any water-quality constituent was for fecal coliform bacteria load at subwatershed 35 (table 14). The largest percentage increases in suspended sediment, dissolved nitrite plus nitrate, and dissolved orthophosphate occurred at Yount Creek (subwatershed 31; 80.0, 95.0, and 50.0 percent, respectively), and the largest percentage increase in dissolved ammonia occurred at Bringle Creek (subwatershed 20; 33.9 percent). The largest percentage increase of total organic carbon and total phosphorus (38.3 and 37.0 percent, respectively) occurred at subwatershed 53, located on the south side of the lake (table 14). As for the entire watershed load, dissolved nitrate plus nitrate and dissolved ammonia increased 8.5 and 11.4 percent, respectively, in scenario 2 relative to the baseline condition, while dissolved orthophosphate and total phosphorous increased 25.0 and 13.8 percent, respectively (table 14). Therefore, phosphorus concentrations increased by greater percentages relative to nitrogen concentrations.

The Lake Maumelle watershed mean streamflow changed very little in response to land-use changes of scenario 2. The simulated mean streamflow into Lake Maumelle increased 7.1 percent from baseline conditions (fig. 30; table 14).

The simulated changes in nutrient, suspended sediment, total organic carbon, and fecal coliform loads from the Lake Maumelle watershed simulated in scenario 2 (fig. 30; table 14) resulted in slight changes in simulated water quality for Lake Maumelle, relative to the baseline condition for both the epilimnion and hypolimnion, for most of the reservoir (fig. 31a–b; table 15). On average, over the simulation period (2004–10), nitrogen concentrations were slightly lower (average total nitrogen for all three sites was 0.01 mg/L lower) than the baseline condition in Lake Maumelle, while dissolved orthophosphate concentration was slightly higher (average for all three sites was 0.001 mg/L higher). Chlorophyll *a* concentrations, on average, over the simulation period (2004–10) were also slightly lower (average for all three sites was 0.1 µg/L lower) than the baseline condition. These differences in total nitrogen, dissolved orthophosphate, and chlorophyll *a* are approximately an order of magnitude less than the RMSE and MAE between measured and predicted concentrations in the baseline model. During the summer season in 2006, both nitrogen and phosphorus concentrations were lower in scenario 2 than in the baseline condition (fig. 34). As a result, chlorophyll *a* concentrations decreased during this same summer season period. The decrease in nitrogen and chlorophyll *a* concentrations during the dry summer season in 2006 was enough to decrease these concentrations, on average (total nitrogen, 0.01 mg/L lower; chlorophyll *a*, 0.1 µg/L lower), over the simulation period (2004–10; fig. 31a).

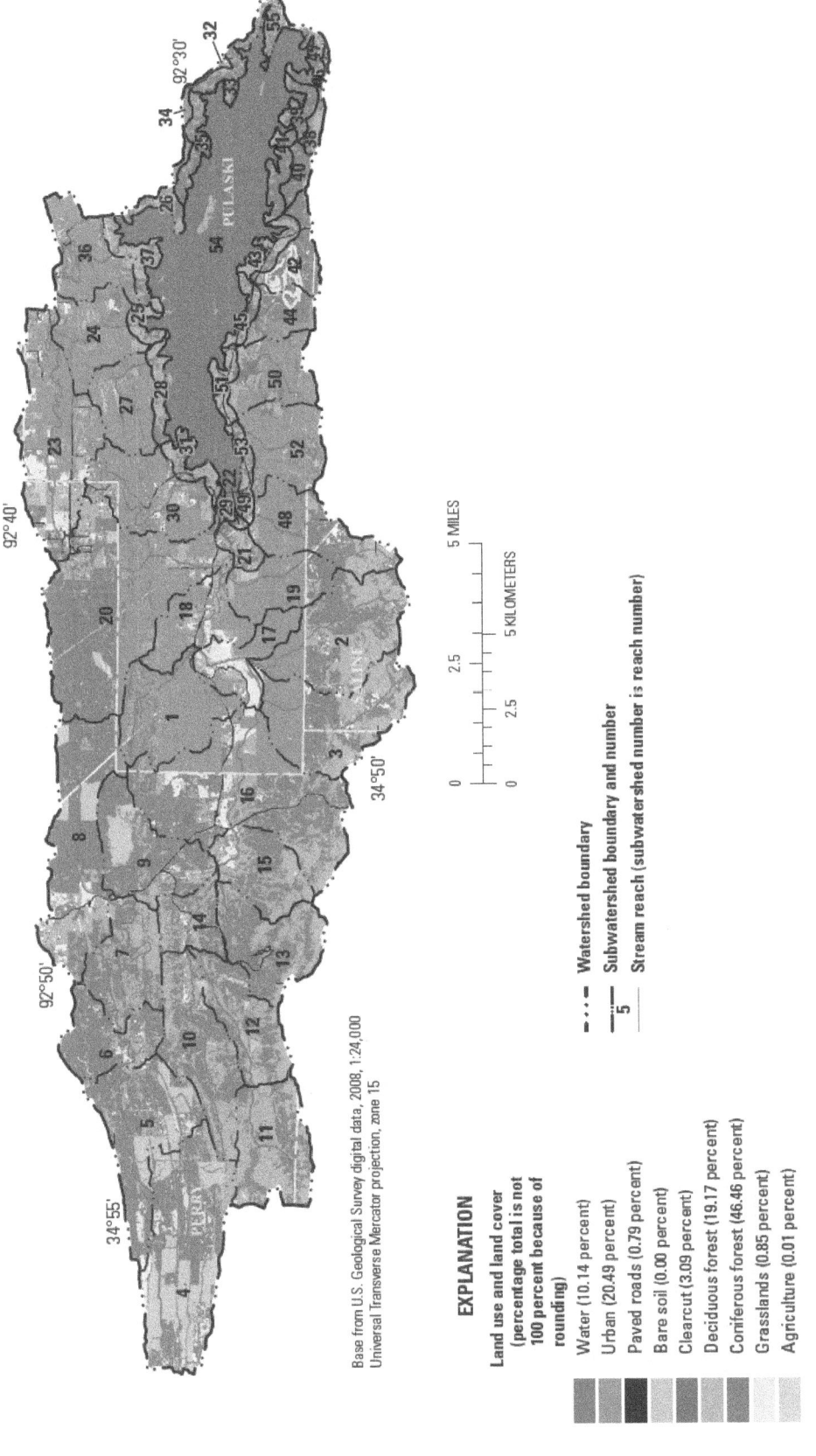

Base from U.S. Geological Survey digital data, 2008, 1:24,000
Universal Transverse Mercator projection, zone 15

EXPLANATION

Land use and land cover
(percentage total is not
100 percent because of
rounding)

Water (10.14 percent)

Urban (20.49 percent)

Paved roads (0.79 percent)

Bare soil (0.00 percent)

Clearcut (3.09 percent)

Deciduous forest (19.17 percent)

Coniferous forest (46.46 percent)

Grasslands (0.85 percent)

Agriculture (0.01 percent)

– · – · – Watershed boundary

––·––·– Subwatershed boundary and number
5

——5 Stream reach (subwatershed number is reach number)

0   2.5   5 KILOMETERS

0   2.5   5 MILES

Figure 33. Low-intensity urban land use for scenario 2.

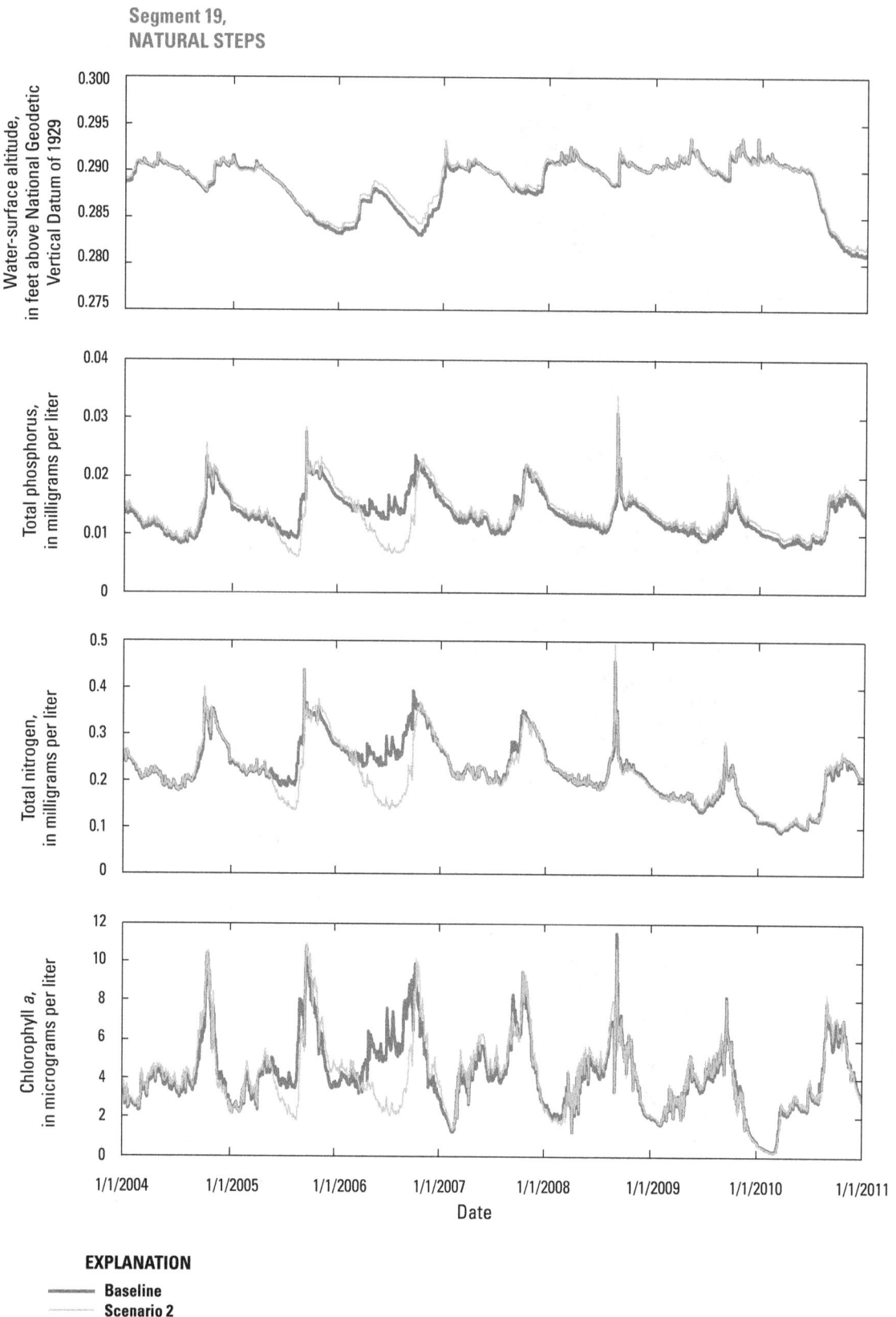

Figure 34. Time-series plots of water-surface altitude, total phosphorus, total nitrogen, and chlorophyll *a* concentrations from scenario 2 results compared against the baseline condition.

The ratio of phosphorus to nitrogen nutrient loading resulting from scenario 2 was disproportionate from that of the baseline condition—the increase in phosphorus was proportionally higher than the increase in nitrogen. The precipitation (containing nitrate and ammonia) on the pool contributes a large percentage of the annual nitrogen load to Lake Maumelle (estimated at 75 percent for the simulation period [2004–10]). During the dry years, 2005 and 2006, the nitrogen load in general was lower than it was in other wetter years. The lower nitrogen loading from precipitation and proportionally larger quantity of phosphorus being loaded off the landscape relative to nitrogen caused primary production (algae growth measured by chlorophyll *a* concentrations) in Lake Maumelle to become limited by the low concentrations of nitrogen earlier in the summer season and across more of the water body than in the baseline condition. Nitrogen limitation was infrequent in the baseline condition. As a result, the lower nitrogen and chlorophyll *a* concentrations during the dry summer season in 2006 (fig. 34) were enough to decrease these concentrations, on average, over the simulation period (2004–10; fig. 31*a*). In the remainder of the simulation period (late 2006–10), phosphorus, nitrogen, and chlorophyll *a* concentrations in Scenario 2 were similar to the baseline condition.

Simulated mean dissolved and total organic carbon concentrations for scenario 2 were slightly higher than the baseline condition concentrations (fig. 31*a–b;* table 15). Simulated mean fecal coliform bacteria densities increased the greatest in the upstream part of the lake and changed slightly in other parts of the epilimnion of the lake for the scenario (fig. 31*a*). Secchi disk depth changed very little in scenario 2 from that of the baseline condition; the greatest change occurred at the upper end of the lake (East of Highway 10) (fig. 31*a–b;* table 15).

Scenario 3—Simulated Clearcutting

Scenario 3 simulated clearcutting within selected forest land in the Lake Maumelle watershed and the addition of roads for timber management and other uses. The selected areas for clearcutting were based on a land-use/land-cover map developed by CAW from aerial photographs that were taken in 2010 (Vince Guillet, Central Arkansas Water, written commun., 2010) around Lake Maumelle. This 2010 land-use/land-cover map that extends only around Lake Maumelle is the most current (at the time of writing this report, 2012) land-use classification for Lake Maumelle and includes new forest clearcuts located around Lake Maumelle in Pulaski County identified by CAW (Vince Guillet, Central Arkansas Water, written commun., 2010) (fig. 35). Additional clearcuts were added to the simulation from Google Earth imagery dated October 14, 2011. Converting the urban, paved roads, grassland, agriculture, and forest land use to clearcuts increased the percentage of clearcuts in the watershed by approximately 3.5 percent (approximately 3,092 acres) of the total acreage (86,907 acres). Roads associated with timber harvesting activities were lumped into the clearcut land-use classification. The acreage for the clearcut areas was added to the appropriate subwatersheds in the HSPF model as clearcut land use, with the equivalent amount subtracted from the forest land use with the subtracted percentage of deciduous and coniferous equal to the 2006 Arkansas land-use/land-cover map (Arkansas Natural Resources Commission and University of Arkansas: Center for Advanced Spatial Technologies, 2009) proportions. Adding in the clearcuts decreased the percentage of forested land in the watershed by approximately 3.5 percent (3,092 acres) of the total acreage (86,907 acres) for the watershed.

An increase of nutrient and sediment movement to streams has been associated with timber management activities (Scoles and others, 2001). The subwatersheds that received clearcuts were 1, 2, 3, 4, 8, 9, 16, 18, 19, 20, 44, 48, 50, and 52 (fig. 35).

The small amount of land-use change, from forested to clearcut, had little overall effect on the water quality within the HSPF model (fig. 30; table 14). Simulated water-quality loading rates transported into Lake Maumelle in scenario 3 were very similar to the baseline condition. Loading rate percentage differences for the Lake Maumelle watershed, as a whole, ranged from 0.1 percent (fecal coliform bacteria) to 4.7 percent (suspended sediment). Percentage differences for individual subwatershed loads ranged from -0.7 percent (dissolved ammonia nitrogen) to 46.7 percent (suspended sediment) (table 14).

The Lake Maumelle watershed mean streamflow changed very little in response to land-use changes of scenario 3. The 2004–10 simulated mean streamflow into Lake Maumelle increased 0.5 percent from baseline conditions (table 14).

The changes in simulated nutrient, suspended sediment, total organic carbon, and fecal coliform bacteria loads from Lake Maumelle watershed for both the epilimnion and hypolimnion from scenario 3 (fig. 31*a–b*; table 14) resulted overall in very slight changes in simulated water quality within Lake Maumelle, relative to the baseline condition, for most of the reservoir (fig. 31*a–b*; table 15). Also, the magnitude of these changes was not substantially different in the upstream, middle, and downstream parts of the reservoir.

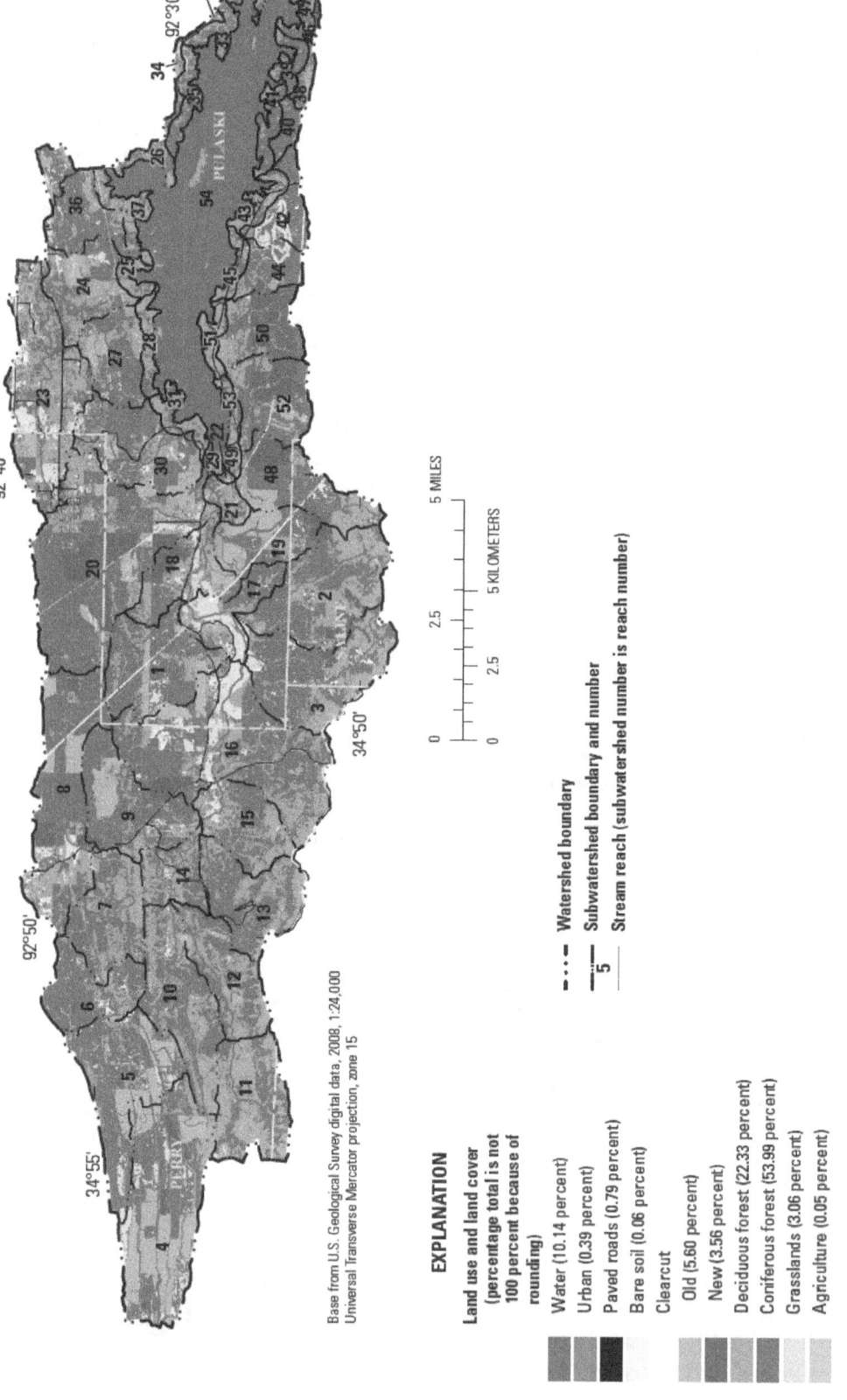

Base from U.S. Geological Survey digital data, 2008, 1:24,000
Universal Transverse Mercator projection, zone 15

EXPLANATION

Land use and land cover
(percentage total is not
100 percent because of
rounding)

Water (10.14 percent)

Urban (0.39 percent)

Paved roads (0.79 percent)

Bare soil (0.06 percent)

Clearcut

Old (5.60 percent)

New (3.56 percent)

Deciduous forest (22.33 percent)

Coniferous forest (53.99 percent)

Grasslands (3.06 percent)

Agriculture (0.05 percent)

– · · ·   Watershed boundary

———   Subwatershed boundary and number

—5—   Stream reach (subwatershed number is reach number)

Figure 35.   Clearcut areas and other land uses for scenario 3, Maumelle Lake, Arkansas.

Implications for Future Monitoring and Management

Through the construction, calibration, and application of the HSPF and CE–QUAL–W2 models, a great deal was learned about the characteristics of the Lake Maumelle watershed and reservoir and the nature of the hydrology and water quality (nutrients and algal production). Future improvements to these and other models will rely on an increased understanding of the complex water-quality processes in the Lake Maumelle watershed and reservoir.

This study provides the unusual case in which water-quality conditions are described and models are used to simulate a watershed and reservoir ecosystem existing in its (most) natural, nonimpaired condition, a condition to which an impaired watershed and reservoir ecosystem would want to be restored, one that requires protection rather than restoration. Nutrient (nitrogen and phosphorus) concentrations in the watershed and reservoir are naturally low and as a result, algal biomass (measured by chlorophyll *a*) is low; the reservoir is in equilibrium with its watershed.

The eutrophication process over the life of Lake Maumelle (1958–2012) has followed the typical conceptual reservoir aging model, where after the reservoir was initially filled, internal nutrient loading resulted from decomposition of inundated vegetation and nutrients released from the inundated soils. These (internally loaded) nutrients provided the resources for increasing primary production (algal production, trophic upsurge) for a limited amount of time (fig. 3), until these internal sources were used up, stored, or exported out of the watershed. After a period of time, following the peak of trophic production (trophic decline) and prior to the model simulation period (2004–10), Lake Maumelle reached equilibrium with the external nutrient loading provided by sources on the landscape within the watershed, demonstrating the rate of aging or eutrophication under "natural" controls. However, with the potential for future human-induced alterations of the watershed, the eutrophication rate will likely increase.

The lack of strong upward trends in the observed Lake Maumelle water quality and in the simulated water quality resulting from the land-use changes in scenarios 1, 2, and 3 indicates that it is important to maintain a consistent ambient water-quality monitoring program in Lake Maumelle. USGS in cooperation with CAW has been monitoring water quality in Lake Maumelle since the summer of 1989. Continued monitoring will provide data and information to determine if water quality in Lake Maumelle is changing or remaining the same. Once, for example, total phosphorus concentrations reach the 0.020 to 0.026 mg/L threshold, one can expect algal production measured by chlorophyll *a* to increase more rapidly with small increases in total phosphorus. Presently, Lake Maumelle remains below this threshold; it is not known if and when the lake will reach this threshold. If the lake was to reach this threshold, it would be prudent to recalibrate the HSPF and

CE–QUAL–W2 models used in this report. The new data will allow the model to simulate changes associated with land use with much more certainty than the present model.

Many of the results described in this report have implications for future management. Among the implications of the results of the modeling described in this report are implications related to spatial scales, location of land-use changes, effects of land use on loading rates, and effects of simulated land-use changes on water quality of Lake Maumelle. Temporally, the magnitude of the water-quality changes simulated by the land-use change scenarios over the 7-year period (2004–10) are not necessarily indicative of the changes that would be expected to occur with similar land-use changes persisting over, for example, a 20-, 30-, 40-year period. Other implications are a direct result of the model limitations, including those related to the aging and eutrophication status of Lake Maumelle. All implications should be tempered by realization of the described model limitations.

To understand the limitations of this specific application of the HSPF and CE–QUAL–W2 models for the Lake Maumelle watershed and receiving reservoir, one must understand the aging and evolution of impounded waterbodies formed by damming and permanently flooding the river valley (fig. 3) as described in "Description of Lake Maumelle Aging and Trophic Status." The relatively unchanging eutrophication response variables for Lake Maumelle indicate that the lake is in equilibrium with its watershed. Model calibration of a lake or reservoir system in equilibrium with its watershed is not difficult by itself, but running simulations or increasing input concentrations to examine response relations can be challenging. Models are calibrated to measured (existing) conditions. If the existing conditions have a homogeneous set of landscape conditions (for example, land use) and a tight range in concentrations (that is they do not vary much) at low concentrations, then selecting values for model parameters (tables 4, 7, and 9) is more difficult because of difficulties in evaluating the appropriateness of a specific parameter value from a range of literature and calibration derived values, as well as difficulties in evaluating the model response.

The accuracy of the HSPF and CE–QUAL–W2 models is limited by the simplification of complexities within the watershed and by data availability (see "Hydrologic Simulation Program–FORTRAN and CE–QUAL–W2 Model Limitations" sections for more details). Comparisons of measured and simulated water-quality characteristics using multiple evaluation methods indicate the accuracy of the model. For example, comparisons between paired observed and HSPF simulated values for various water-quality measures ranged from poor to good based on percentage differences. However, large percentage differences could result from relatively small absolute differences (for example, a 44-percent difference in dissolved nitrite plus nitrate concentrations associated with an absolute difference of less than 0.3 mg/L). Because suspended-sediment concentrations are one of the more uncertain model outputs and one of the most difficult water-quality constituents to accurately represent in current watershed models (U.S. Environmental Protection Agency, 2006), this is one of the model results that should be considered more skeptically than others.

Use of a watershed model (HSPF) and a reservoir model (CE–QUAL–W2) of the Lake Maumelle watershed to compare baseline conditions with three scenarios to simulate changes in land use indicates that the simulated changes in land use resulted in little change in the water quality within the lake (see scenario discussions within the "Effects of Simulated Land-Use Changes on Water Quality" section for quantifications of the changes). Simulated land-use changes affected approximately 9.9 percent (scenario 1), 20 percent (scenario 2), and 3.5 percent (scenario 3) of the Lake Maumelle watershed. The largest simulated changes in water quality generally occurred as a result of the land-use changes simulated in scenario 2. Land-use changes simulated in scenario 2 affected the largest area (approximately 20 percent of the watershed) and generally were closer to Lake Maumelle than the land-use changes simulated in the other scenarios (figs. 29, 33, and 35). This indicates that the number of acres affected by a land-use change and the proximity to Lake Maumelle (and associated transport pathways) are important factors in determining the effect of land-use change on water quality of Lake Maumelle. A third factor is the intensity (determined by the amount of disruption of existing processes) of the land-use change.

The simulated loading rates for total phosphorus, total organic carbon, and suspended sediment from subwatersheds in the Lake Maumelle watershed (which are subject to the same uncertainties as the simulated concentrations of total phosphorus, total organic carbon, and suspended sediment) frequently were less than the values for performance standards (also termed "site-scale pollution allocations" or "pollution loading limits") for Critical Area A, Critical Area B, and the UWA from the watershed management plan (Tetra Tech, Inc., 2007) (fig. 1). However, these performance standards from the watershed management plan cannot validly be compared with the loading rates for the baseline condition or for the simulations because the watershed management plan standards are values for site-scale (or upland) loads rather than values for loads delivered to the lake (Tetra Tech Inc., memorandum of December 18, 2007, in Arkansas Department of Environmental Quality, 2007).

As mentioned previously in the description of model limitations, changes in simulated water quality resulting from these scenarios (or other scenarios) do not necessarily have any relation to water-quality changes that might occur after conditions persist for a longer time period than was modeled. Indeed, because the model results are partly the result of weather conditions (rainfall, temperature, wind conditions) the model results cannot be extended to any other 7-year period without being cognizant of the effects of weather conditions. This means that the magnitude of the water-quality changes simulated by the scenarios over the 7-year period (2004–10) are not necessarily indicative of the changes that could be

expected to occur with similar land-use changes persisting over, for example, a 20-, 30-, or 40-year period.

Summary

Lake Maumelle, located in central Arkansas northwest of the cities of Little Rock and North Little Rock, is one of two principal drinking-water supplies for the Little Rock and North Little Rock metropolitan areas. The drainage area upstream from the spillway is approximately 137 square miles. Approximately 80 percent of the land area within the entire Lake Maumelle watershed is forest, approximately 10 percent water (including Lake Maumelle), approximately 5.6 percent clearcut area, and approximately 3 percent grasslands. Lake Maumelle and the Maumelle River are more pristine than most other reservoirs and streams in the region. However, as the Lake Maumelle watershed becomes more urbanized and timber harvesting becomes more extensive, concerns about the sustainability of the quality of the water supply also have increased.

Given the reservoir aging process described in this report, Lake Maumelle has since passed the trophic disequilibrium phases resulting from impoundment ("trophic upsurge" and following "trophic depression") and currently coexists in equilibria with its watershed (external loading) where productivity has remained relatively constant over time. Undisturbed terrestrial ecosystems are usually characterized by runoff with low concentrations in dissolved substances; however, pastures, croplands, and urban areas contribute much greater nutrient loads to aquatic systems. Therefore, land-use patterns will have long-term effects on reservoir productivity and water quality. If reservoirs are permitted to age without being otherwise disturbed, one would expect (based on present understanding of the relations between basin morphology, nutrient loading rates, and lacustrine productivity) that reservoir productivity would either remain relatively constant over time (for reservoirs that fill slowly) or gradually increase as mean depth decreases (for reservoir basins undergoing rapid siltation). Because construction of reservoirs (manmade impoundments) often promote additional land-use changes and technological development within reservoir watersheds and their relatively large watersheds focus both point and diffuse sources of nutrients into reservoir basins, water quality and productivity changes attributable to "natural" reservoir aging will be small compared to the effects of human-induced changes in watershed-reservoir interactions.

Two hydrodynamic and water-quality models were developed by the U.S. Geological Survey, in cooperation with Central Arkansas Water, to partially address these concerns. A Hydrologic Simulation Program–FORTRAN (HSPF) watershed model was developed to simulate streamflow, water temperature, dissolved oxygen, suspended sediment, total organic carbon, dissolved ammonia nitrogen, dissolved nitrite plus nitrate nitrogen, dissolved orthophosphate, total phosphorus, and fecal coliform bacteria using input data collected from January 2004 through 2010. A CE–QUAL–W2 model was developed to simulate reservoir hydrodynamics and selected water-quality characteristics including temperature, fecal-coliform bacteria concentrations, nutrient concentrations, organic-carbon concentrations, algae groups, and chlorophyll *a* concentrations using the simulated output from the HSPF model from January 2004 through 2010.

Numerous datasets were required in the development of the HSPF model. Datasets compiled for this study include the National Elevation Dataset for use in determining hydrologically similar land areas; the National Hydrography Dataset that includes all the stream reaches within the watershed; the 2006 Arkansas land-use/land-cover maps; aerial photography taken February 2009; the Soil Survey Geographic database for each county within the watershed; and Next-Generation Radar hourly precipitation data as well as other meteorological data including air temperature, solar radiation, dew point temperature, wind velocity, and cloud cover. Air temperature, dewpoint temperature, wind velocity, and cloud cover were obtained from the National Climatic Data Center stations surrounding the watershed.

The HSPF watershed model was calibrated to five streamflow-gaging stations, and in general, these stations characterize a range of subwatershed areas with varying land-use types. Continuous streamflow data, discrete sediment concentration data, and other discrete water-quality data were used to calibrate the Lake Maumelle HSPF model. The HSPF model was developed using 55 subwatersheds, and therefore, 55 stream reaches to characterize Lake Maumelle's watershed. The simulated watershed area covers approximately 136 mi^2 with eight pervious land types and seven impervious land types and approximately 80 percent of the entire watershed is classified within the forest land use.

The CE–QUAL–W2 reservoir model was calibrated to water-quality data collected during January 2004 through December 2010 at three reservoir stations. The CE–QUAL–W2 model simulates 14 active and 6 derived constituents and hydraulic, thermal, and chemical boundary conditions were required. Development of the CE–QUAL–W2 model of Lake Maumelle included the computational grid, specification of boundary and initial conditions, and preliminary selection of model parameter values. The boundaries of the Lake Maumelle model included the reservoir bottom, the shoreline, tributary streams, the upstream boundary, the downstream boundary, and the water-surface altitude of the reservoir.

In general, the baseline simulation for the HSPF and CE–QUAL–W2 models matched reasonably well to the measured data. In general, based on the exceedance probability, simulated "low flows" (in this instance, flows with exceedance probabilities greater than about 60 to 70 percent) were greater than the measured low flows for three inflow stations, but simulated high flows matched reasonably well to observed high flows. Streamflow calibration results were in close agreement at both high and low flows for one additional inflow station, whereas simulated low flows were less than the measured low flows, but simulated high flows matched reasonably well to observed high flows at a fifth inflow site.

In general, simulated and measured suspended-sediment concentrations during periods of base flow (streamflows not substantially influenced by runoff) agreed reasonably well for inflow stations with differences—simulated minus measured value—(80 percent of the values) ranging from -15 to 41 mg/L, and percent difference—relative to the measured value—generally ranging from -99 to 182 percent at one inflow site and -20 to 22 mg/L, (-100 to 194 percent) at another. Additionally, simulated suspended-sediment concentrations matched well with the quarterly and monthly sampling values and also, during periods of stormflow (streamflow substantially influenced by runoff). Generally, this also was the case for fecal coliform bacteria numbers and total organic carbon and nutrient concentrations. In general, water temperature and dissolved-oxygen concentration simulations followed measured seasonal trends for all stations with the largest differences occurring during periods of lowest temperatures (for temperature) or during the periods of lowest measured dissolved-oxygen concentrations (for dissolved oxygen).

For the CE–QUAL–W2 model, simulated vertical distributions of temperatures agreed with measured distributions even for complex temperature profiles. Although the calibrated model generally provided an excellent simulation of water temperature in Lake Maumelle, the simulation accuracy of water temperatures varied with water temperature season and depth. The onset of low dissolved-oxygen concentrations and the recovery to higher dissolved-oxygen concentrations were well simulated throughout the reservoir. Considering the oligotrophic-mesotrophic (low to intermediate primary productivity and associated low nutrient concentrations) condition of Lake Maumelle, simulated algae (generally within 2 to 3 µg/L as chlorophyll *a*), phosphorus (generally within 0.01 to 0.02 mg/L), and nitrogen (generally within 0.1 to 0.2 mg/L) concentrations generally compared well with measured values. Simulated fecal-coliform bacteria concentrations for Lake Maumelle exhibited the same general patterns and magnitudes as measured values.

The calibrated HSPF model and the calibrated CE–QUAL–W2 model were developed to simulate three land-use scenarios and ascertain the potential effects of these land-use changes on the water quality of Lake Maumelle. These scenarios included a scenario that simulated conversion of most land in the watershed to forest (scenario 1), a scenario that simulated conversion to low-intensity urban land use in part of the watershed (scenario 2), and a scenario that simulated timber harvest in part of the watershed (scenario 3). Simulated land-use changes for scenarios 1 and 3 resulted in little overall effect on the simulated water quality in the HSPF model. The land-use change of scenario 2 affected most subwatersheds that drain into the northern side of Lake Maumelle and resulted in large percentage increases (generally between 10 and 25 percent) in dissolved nitrite plus nitrate nitrogen, dissolved ammonia nitrogen, dissolved orthophosphate, total phosphorus, total organic carbon, fecal coliform bacteria, and suspended sediment loading rates.

For scenario 1, the simulated changes in nutrient, suspended sediment, total organic carbon, and fecal coliform bacteria loads from the HSPF model resulted in very slight changes in simulated water quality for Lake Maumelle, relative to the baseline condition. Following lake mixing in the fall of 2006 and 2007, phosphorus and nitrogen concentrations were higher than the baseline condition and chlorophyll *a* responded accordingly. The increased nutrient and chlorophyll *a* concentrations in late October and into 2007 were enough to increase concentrations, on average, over the entire simulation period (2004–10).

For scenario 2, the simulated changes in nutrient, suspended sediment, total organic carbon, and fecal coliform bacteria loads from the Lake Maumelle watershed resulted in slight changes in simulated water quality for Lake Maumelle, relative to the baseline condition (total nitrogen decreased by 0.01 milligram per liter; dissolved orthophosphate increased by 0.001 milligram per liter; chlorophyll *a* decreased by 0.1 microgram per liter). The differences in these concentrations are approximately an order of magnitude less than the error between measured and simulated concentrations in the baseline model. During the driest summer in the simulation period (2006), phosphorus and nitrogen concentrations were lower than the baseline condition and chlorophyll *a* concentrations decreased during the same summer season. The decrease in nitrogen and chlorophyll *a* concentrations during the summer in 2006 was enough to decrease concentrations of these constituents very slightly, on average for the entire simulation period.

For scenario 3, the changes in simulated nutrient, suspended sediment, total organic carbon, and fecal coliform bacteria loads from Lake Maumelle watershed resulted overall in very slight changes in simulated water quality within Lake Maumelle, relative to the baseline condition, for most of the reservoir.

Among the implications of the results of the modeling described in this report are implications related to spatial scales and location of land-use changes, effects of land use on loading rates, and effects of simulated land-use changes on water quality of Lake Maumelle. Temporally, the magnitude of the water-quality changes simulated by the land-use change scenarios over the 7-year period (2004–10) are not necessarily indicative of the changes that could be expected to occur with similar land-use changes persisting over a 20-, 30-, or 40-year period, for example. These implications should be tempered by realization of the described model limitations.

References Cited

Al-Abed, N.A., and Whiteley, H.R., 2002, Calibration of the Hydrologic Simulation Program Fortran (HSPF) model using automatic calibration and geographical information systems: Hydrological Processes, v. 16, p. 3169–3188.

Arkansas Department of Environmental Quality, 2007, Technical basis for the Lake Maumelle watershed plan recommendation for prohibiting point source discharges: Tetra Tech memorandum dated December 18, 2007.

Arkansas Department of Environmental Quality, 2008, Integrated monitoring and assessment report: Arkansas Department of Environmental Quality, variously paginated.

Arkansas Natural Resources Commission and University of Arkansas: Center for Advanced Spatial Technologies, 2009, 2006 Arkansas land use/land cover, accessed March 13, 2009, at http:// www.cast.uark.edu/home/research/ geomatics/remote-sensing/arkansas-land-useland-cover/2006-arkansas-land-use-land-cover-project html/.

Bales, J.D., Sarver, K.M., and Giorgino, M.J., 2001, Mountain Island Lake, North Carolina: Analysis of ambient conditions and simulation of hydrodynamics, constituent transport, and water-quality characteristics, 1996–97: U.S. Geological Survey Water-Resources Investigations Report 01–4138, 85 p.

Baranov, I.V., 1961, Biohydrochemical classification of reservoirs in the European U.S.S.R., *in* Tyurin, P.V., ed., The storage lakes of the U.S.S.R. and their importance for fishery: Israel Program for Scientific Translations, Tel Aviv, Israel, p. 139–183.

Baxter, R.M., 1977, Environmental effects of dams and impoundments: Annual Review of Ecology and Systematics v. 8, p. 255–283.

Benson, N.G., 1982, Some observations on the ecology and fish management of reservoirs in the United States: Canadian Water Resources Journal, v. 7, p. 2–25.

Bicknell, B.R., Imhoff, J.C., Kittle, J.L., Jr., Jobes, T.H., Donigian, A.S., Jr., 2001, Hydrologic Simulation Program–FORTRAN, user's manual for Version 12: Research Triangle Park, N.C., U.S. Environmental Protection Agency, National Exposure Research Laboratory, Office of Research and Development, 845 p.

Brossett, T.H, and Evans, D.A., 2003, Water resources data, Arkansas, water year 2002: U.S. Geological Survey Water-Data Report AR 02–1, 461 p.

Brossett, T.H., Schrader, T.P., and Evans, D.A., 2005, Water resources data, Arkansas, water year 2004: U.S. Geological Survey Water-Data Report AR 04–1, 460 p.

Bureau of Land Management, 1983, Environmental impact statement, Glenwood Springs Resource Area Glenwood Springs, Colorado: Department of the Interior, Bureau of Land Management Resource Management Plan, 255 p.

Byron, J. and Burchard, D., 2008, Septic systems permitting in Luna Lake/Alpine – Summer 2008: Arizona Department of Environmental Quality Fact Sheet, 2 p.

Childress, C.J.O., Foreman, W.T., Conner, B.F., Maloney, T.J., 1999, New reporting procedures based on long-term method detection levels and some considerations of water-quality data provided by the U.S. Geological Survey National Water Quality Laboratory: U.S. Geological Survey Open-File Report 99–193, 19 p.

Cohn, T.A., 1988, Adjusted maximum likelihood estimation of the moments of lognormal populations from type 1 censored samples: U.S. Geological Survey Open-File Report 88–350, 34 p.

Cohn, T.A., DeLong, L.L., Gilroy, E.J., Hirsch, R.M., and Wells, D.K., 1989, Estimating constituent loads: Water Resources Research, v. 2, no. 5, p. 937–942.

Cohn, T.A., Gilroy, E.J., and Baier, W.G., 1992, Estimating fluvial transport of trace constituents using a regression model with data subject to censoring: Proceedings of the Joint Statistical Meeting, Boston, August 9–13, 1992, p. 142–151.

Cole, T.M., and Wells, S.A., 2008, CE–QUAL–W2: A two-dimensional, laterally averaged, hydrodynamic and water quality model, version 3.6: U.S. Army Corps of Engineers Instruction Report EL–08–1, variously paginated.

Crawford, C.G., 1991, Estimation of suspended-sediment rating curves and mean suspended-sediment loads: Journal of Hydrology, v. 129, p. 331–348.

Dempster, A.P., Laird, N.M., and Rubin, D.B., 1977, Maximum likelihood from incomplete data via the EM algorithm: Journal of the Royal Statistical Society, Series B, v. 39, no. 1, p. 1–38.

Dillon, P.J., and Rigler, F.M., 1974, The chlorophyll-phosphorus relationship in lakes: Limnology and Oceanography, v. 19, p. 767–773.

Donigian, A.S., Jr., 2000, HSPF Training Workshop Handbook and CD, Lecture #19—Calibration and Verification Issues, Slide #L19–22: Environmental Protection Agency Headquarters, Washington Information Center, 10–14 January, 2000, Presented and prepared for U.S. EPA, Office of Water, Office of Science and Technology, Washington, D.C.

Donigian, A.S., Jr., and Crawford, N.H., 1977, Simulation of nutrient loadings in surface runoff with the NPS model: U.S. Environmental Protection Agency Ecological Research Series EPA–600/3–77–065, 110 p.

Donigian, A.S., Jr., and Davis, H.H., 1978, User's manual for Agricultural Runoff Management (ARM) model: U.S. Environmental Protection Agency Report EPA–600/3–78–080, 163 p.

Evans, D.A., Brossett, T.H., and Schrader, T.P., 2004, Water resources data, Arkansas, water year 2003: U.S. Geological Survey Water-Data Report AR 03–1, 450 p.

Evans, D.A., Porter, J.E., and Westerfield, P.W., 1995, Water resources data, Arkansas, water year 1994: U.S. Geological Survey Water-Data Report AR 94–1, 466 p.

Fenneman, N.M., and Johnson, D.W., 1946, Physical divisions of the United States (Map): Washington, D.C., U.S. Geological Survey.

Gaffney, J.S., and Marley, N.A., 2010, Anthropogenic perturbations of biogenic aerosols—Climate impacts and feedbacks: 12th Conference on Atmospheric Chemistry, 2nd Cymposium on Aerosol-Cloud-Climate Interactions, Joint Session 17, American Meteorology Society, accessed March 21, 2012, at https://ams.confex.com/90annual/techprogram/paper_164772 htm.

Galloway, J.M., and Green, W.R., 2002, Simulation of hydrodynamics, temperature, and dissolved oxygen in Norfork Lake, Arkansas, 1994–95: U.S. Geological Survey Water-Resources Investigations Report 02–4250, 23 p.

Galloway, J.M., and Grccn, W.R., 2003, Simulation of hydrodynamics, temperature, and dissolved oxygen in Bull Shoals Lake, Arkansas, 1994–1995: U.S. Geological Survey Water-Resources Investigations Report 03–4077, 23 p.

Galloway, J.M., and Green, W.R., 2004, Water-quality assessment of Lakes Maumelle and Winona, Arkansas, 1991 through 2003: U.S. Geological Survey Scientific Investigations Report 2004–5182, 46 p.

Gilroy, E.J., Hirsch, R.M., and Cohn, T.A., 1990, Mean square error of regression-based constituent transport estimates: Water Resources Research, v. 26, no. 9, p. 2069–2077.

Gloss, S.P., Mayer, L.M., and Kidd, D.E., 1980, Advective control of nutrient dynamics in the epilimnion of a large reservoir: Limnology and Oceanography, v. 25, p. 219–228.

Google Earth, 2011, Imagery of the Lake Maumelle watershed, accessed on January 4, 2012, at http://www.google.com/earth/index html.

Green, W.R., 1994, Water quality assessment of Maumelle and Winona reservoir systems, central Arkansas, May 1989–October 1992: U.S. Geological Survey Water-Resources Investigations Report 93–4218, 42 p.

Green, W.R., 1998, Water-quality assessment of the Frank Lyon, Jr., nursery pond releases into Lake Maumelle, Arkansas, 1991–1996: U.S. Geological Survey Water-Resources Investigations Report 98–4194, 38 p.

Green, W.R., 2001, Analysis of ambient conditions and simulation of hydrodynamics, constituent transport, and water-quality characteristics in Lake Maumelle, Arkansas, 1991–92: U.S. Geological Survey Water-Resources Investigations Report 01–4045, 60 p.

Green, W.R., and Louthian, B.L., 1993, Hydrologic data collected in Maumelle and Winona reservoir systems, central Arkansas, May 1989 through October 1992: U.S. Geological Survey Open-File Report 93–122, 253 p.

Green, W.R., Galloway, J.M., Richards, J.M., and Wesolowski, E.A., 2003, Simulation of hydrodynamics, temperature, and dissolved oxygen in Table Rock Lake, Missouri, 1996–1997: U.S. Geological Survey Water-Resources Investigations Report 03–4237, 35 p.

Greene, D.G., and Hudlow, M.D., 1982, Hydrometeorologic grid mapping procedures, *in* Johnson, A.I., and Clark, R.A., American Water Resources Association International Symposium on Hydrometeorology, Proceedings: Denver, Colorado, June 13–17, 1982, 20 p.

Haggard, B.E., and Green, W.R., 2002, Simulation of hydrodynamics, temperature, and dissolved-oxygen in Beaver Lake, Arkansas, 1994–1995: U.S. Geological Survey Water-Resources Investigations Report 02–4116, 21 p.

Haggard, B.E., DeLaune, P.B., Smith, D.R., and Moore, P.A., Jr., 2005, Nutrient and β_{17}-Estradiol loss in runoff water from poultry litters: Journal of the American Water Resources Association, v. 41, issue 2, p. 245–256.

Heathman, G.C., Sharpley, A.N., Smith, S.J., and Robinson, J.S., 1995, Land application of poultry litter and water quality in Oklahoma, U.S.A.: Fertilizer Research, v. 40, p. 165–173.

Helsel, D.R., and Hirsch, R.M., 2002, Statistical methods in water resources: U. S. Geological Survey Techniques of Water-Resource Investigations, book 4, chap. A3, 523 p.

Hummel, P., Kittle, J., and Gray, M., 2001, WDMUtil, Version 2.0: A tool for managing watershed modeling time-series Data: U.S. Environmental Protection Agency Contract No. 68-C-98-010, 157 p.

Hutchinson, G.E., 1973, Eutrophication: American Scientist, v. 6, p. 269–279.

Kimmel, B.L., and Groeger, A.W., 1986, Limnological and ecological changes associated with reservoir aging, *in* Hall, G.E., and VanDen Avyle, M.J., eds., Reservoir fisheries management—Strategies for the 80's: Bethesda, Maryland, Southern Division American Fisheries Society, p. 103–109.

Kimmel, B.L., Lind, O.T., and Paulson, L.J., 1990, Reservoir primary production, *in* Thornton, K.W., Kimmel, B.L., and Payne, F.E., eds., Reservoir limnology—Ecological perspectives: New York, John Wiley and Sons, p. 133–194.

Kleinman, P.J.A., and Sharpley, A.N., 2003, Effect of broadcast manure on runoff phosphorus concentrations over successive rainfall events: Journal of Environmental Quality, v. 32, p. 1072–1081.

Likens, G.E., 1972, Eutrophication and aquatic ecosystems, *in* Likens, G.E., ed., Nutrients and eutrophication: the limiting-nutrient controversy: American Society of Limnology and Oceanography Special Symposium 1, Allen Press, Lawrence, Kansas, p. 2–13.

Likens, G.E., 1975, Nutrient flux and cycling in freshwater ecosystems, *in* Howell, F.G., Gentry, J.B., and Smith, M.H., eds., Mineral cycling in southeastern ecosystems: U.S. Energy Research and Development Agency, CONF–740513, Springfield, Virginia, National Technical Information Service.

Likes, Jiri, 1980, Variance of the MVUE for lognormal variance: Technometrics, v. 22, no. 2, p. 253–258.

Maryland Department of the Environment, 2006, Maryland's 2006 TMDL implementation guidance for local governments: The Maryland Department of the Environment, May 24, 2006, 179 p.

Maryland Department of the Environment, 2011, General Assembly 2011: Reducing pollution from septic systems, moving carefully on Marcellus Shale, countering acid mine drainage: eMDE, Quarterly Online Newsletter, v. IV, no. 9, accessed on August 28, 2012, at http://www.mde.state.md.us/programs/ResearchCenter/ReportsandPublications/eMDE/Pages/researchcenter/publications/general/eMDE/vol4no9/Article4.aspx.

Moore, K., and Mohamoud, Y., 2007, Two automated methods for creating hydraulic function tables (FTABLES): U.S. Environmental Protection Agency BASINS Technical Note 2, 14 p.

Morris, E.E., Porter, J.E., and Westerfield, P.W., 1992, Water resources data, Arkansas, water year 1991: U.S. Geological Survey Water-Data Report AR 91–1, 588 p.

Myers, D.N., and Wilde, F.D., 1999, Biological indicators: U.S. Geological Survey Techniques of Water-Resources Investigations, book 9, chap. A7, variously paginated.

National Atmospheric Deposition Program, 2012, NADP/NTN Monitoring Location AR03, accessed January 3, 2012, at http:// nadp.sws.uiuc.edu/sites/siteinfo.asp?id=AR03&net=NTN.

National Climatic Data Center, 2008, Multi-sensor Precipitation Reanalysis, accessed January 26, 2012 at http://www.ncdc.noaa.gov/oa/rsad/mpr html.

National Oceanic and Atmospheric Administration, 2011, NOAA Satellite and Information Service, accessed October 26, 2011, at http://www ncdc noaa.gov/oa/ncdc.html.

National Oceanic and Atmospheric Administration, 2002, About the Stage III data, accessed January 26, 2012, at http://www nws noaa.gov/oh/hrl/dmip/stageiii_info htm.

National Oceanic and Atmospheric Administration, 2008, Archive of River Forecast Center Operational NEXRAD Data: accessed in January 26, 2012, at http://dipper nws.noaa.gov/hdsb/data/nexrad/nexrad html.

National Oceanic and Atmospheric Administration, 2010, ABRFC Precipitation Processing, accessed January 26, 2012, at http://www.srh.noaa.gov/abrfc/?n=pcpn_methods.

National Oceanic and Atmospheric Administration, 2012, U.S. Climate at a Glance, accessed January 9, 2012, at http://www.ncdc.noaa.gov/oa/climate/research/cag3/ar.html.

Ockerman, D.J., and Roussel, M.C., 2009, Simulation of streamflow and water quality in the Leon Creek watershed, Bexar County, Texas, 1997–2004: U.S. Geological Survey Scientific Investigations Report 2009–5191, 50 p.

Ploskey, G.R., 1981, Factors affecting fish production and fishing quality in new reservoirs with guidance on timber clearing, basin preparation, and filling: Vicksburg, Mississippi, U.S. Army Engineer Waterways Experiment Station Technical Report E–81–11.

Pomes, M.L., Green, W.R., Thurman, E.M., Orem, W.H., and Lerch, H.T., 1997, Sources characterization of disinfection byproduct precursors in two Arkansas water-supply reservoirs: U.S. Geological Survey Fact Sheet 118–97, 4 p.

Pomes, M.L., Green, W.R., Thurman, E.M., Orem, W.H., and Lerch, H.T., 1999, DBP formation potential of aquatic humic substances: Journal American Water Works Association, v. 91, no. 3, p. 103–115.

Porter, J.E., Westerfield, P.W., and Morris, E.E., 1993, Water resources data, Arkansas, water year 1992: U.S. Geological Survey Water-Data Report AR 92–1, 461 p.

Porter, J.E., Evans, D.A., and Pugh, A.L., 1996, Water resources data, Arkansas, water year 1995: U.S. Geological Survey Water-Data Report AR 95–1, 408 p.

Porter, J.E., Evans, D.A., and Remsing, L.M., 1997, Water resources data, Arkansas, water year 1996: U.S. Geological Survey Water-Data Report AR 96–1, 238 p.

Porter, J.E., Evans, D.A., and Remsing, L.M., 1998, Water resources data, Arkansas, water year 1997: U.S. Geological Survey Water-Data Report AR 97–1, 288 p.

Porter, J.E., Evans, D.A., and Remsing, L.M., 1999, Water resources data, Arkansas, water year 1998: U.S. Geological Survey Water-Data Report AR 98–1, 358 p.

Porter, J.E., Evans, D.A., and Remsing, L.M., 2000, Water resources data, Arkansas, water year 1999: U.S. Geological Survey Water-Data Report AR 99–1, 321 p.

Porter, J.E., Evans, D.A., and Remsing, L.M., 2001, Water resources data, Arkansas, water year 2000: U.S. Geological Survey Water-Data Report AR 00–1, 403 p.

Porter, J.E., Evans, D.A., and Remsing, L.M., 2002, Water resources data, Arkansas, water year 2001: U.S. Geological Survey Water-Data Report AR 01–1, 396 p.

Prairie, Y.T., Duarte, C.M., and Klaff, J., 1989, Unifying nutrient-chlorophyll relationships in lakes: Canadian Journal of Fisheries and Aquatic Science, v. 46, p. 1176–1182.

Rantz, S.E., and others, 1982, Measurement and computation of streamflow—Volume 1. Measurement of stage and discharge: U.S. Geological Survey Water-Supply Paper 2175, 284 p.

Runkel, R.L., Crawford, C.G., and Cohn, T.A., 2004, Load estimator (LOADEST)—A FORTRAN program for estimating constituent loads in streams and rivers: U.S. Geological Survey Techniques and Methods, book 4, chap. A5, accessed July 2010 at http://pubs.er.usgs.gov/usgspubs/tm/tm4A5.

Ryu, J.H., 2009, Application of HSPF to the distributed model intercomparison project: Case study: Journal of Hydrologic Engineering, v. 14, no. 8, p. 847–857

Schrader, T.P., Evans, D.A., and Brossett, T.H., 2006, Water resources data Arkansas, water year 2005: U.S. Geological Survey Water-Data Report AR 05–1, 437 p.

Scoles, S., Anderson, S., Turton, D., and Miller, E., 2001, Forestry and water quality, A review of watershed research in the Ouachita Mountains: Water-Quality Series Circular E–932, Oklahoma Cooperative Extension Service, Oklahoma State University, Stillwater, Okla., 29 p.

Sharpley, A.N., 1997, Rainfall frequency and nitrogen and phosphorus runoff from soil amended with poultry litter: Journal of Environmental Quality, v. 26, no. 4, p. 1127–1132.

Shedd, R.C., and Fulton, R.A., 1993, WSR–88D Precipitation processing and its use in National Weather Service hydrologic forecasting, in Engineering Hydrology, Proceedings of the Symposium, San Francisco, Calif., July 25–30, 1993: Proceedings, Hydraulics Division of the American Society of Civil Engineers.

Skahill, B.E., 2003, HSPF modeling at the Engineer Research and Development Center: Research Hydraulic Engineer, Watershed Systems Group, Hydrologic Systems Branch, Coastal and Hydraulics Laboratory, U.S. Army Engineer Research Development Center, accessed December 3, 2010, at http://www.hec.usace.army.mil/misc/watershed_conference/PDF_Files/Skahill_Brian.pdf.

Soong, D.T., Straub, T.D., Murphy, E.A., 2005, Continuous hydrologic simulation and flood-frequency, hydraulic, and flood-hazard analysis of the Blackberry Creek Watershed, Kane County, Illinois: U.S. Geological Survey Scientific Investigations Report 2005–5270, 66 p.

Straskraba, Milan, 1976, Empirical and analytical models of eutrophication: Proceedings of European Symposium, Karl-Marx-Stadt, East Germany, v. 3, p. 352–371.

Straskraba, Milan, 1978, Theoretical considerations on eutrophication: International Association of Theoretical and Applied Limnology, v. 20, p. 2714–2720.

Straskraba, Milan, 1985, Managing eutrophication my means of ecothnology and mathematical modeling, *in* Lakes pollution and recovery: International Congress, Rome, April 15–18, 1985, p. 17–28.

Straskraba, Milan, and others, 1993, Framework for investigation and evaluation of reservoir water quality in Czechoslovakia, *in* Straskraba, M., Tundisi, J.G., and Duncan, A., eds., Comparative reservoir limnology and water quality management: Netherlands, Kluwer Academic Publishers, p. 169–212.

Sullivan, A.B., and Rounds, S.A., 2005, Modeling hydrodynamnics, temperature, and water quality in Henry Hagg Lake, Oregon, 2000–03: U.S. Geological Survey Scientific Investigations Report 2004–5261, 38 p.

Tetra Tech, Inc., 2004, Translation of SET loading estimates to Upper Neuse performance targets: Research Triangle Park, N.C., Tetra Tech, Inc., Memorandum dated February 19, 2004.

Tetra Tech, Inc., 2006, Lake Maumelle watershed and lake modeling—Model calibration report: Research Triangle Park, N.C., Tetra Tech, Inc., 120 p.

Tetra Tech, Inc., 2007, Lake Maumelle watershed management plan: Research Triangle Park, N.C., Tetra Tech, Inc., 130 p.

TIBCO Software Inc., 2008, TIBCO Spotfire S+ 8.1 for Windows.

U.S. Army Corps of Engineers, 1994, Flood-Runoff Analysis: U.S. Army Corps of Engineers Engineering Manual, EM 1110–2–1417, 214 p.

U.S. Department of Agriculture, 1975a, Soil survey of Perry County, Arkansas: Soil Conservation Service and Forest Service in cooperation with Arkansas Agricultural Experiment Station, 111 p.

U.S. Department of Agriculture, 1975b, Soil survey of Pulaski County, Arkansas: Soil Conservation Service and Forest Service in cooperation with Arkansas Agricultural Experiment Station, 111 p.

U.S. Department of Agriculture, 1975c, Soil survey of Saline County, Arkansas: Soil Conservation Service and Forest Service in cooperation with Arkansas Agricultural Experiment Station, 111 p.

U.S. Department of Agriculture, 2007, 2007 Census of agriculture, Arkansas state and county data: National Agricultural Statistics Service, Volume 1, Geographic Area Series, Part 4, AC–07–A–4.

U.S. Department of Agriculture, 2009, Soil Survey Geographic (SSURGO) database: National Resources Conservation Service, accessed May 4, 2009, at http://soils.usda.gov/survey/geography/ssurgo/.

U.S. Environmental Protection Agency, 1999, Preliminary data summary of urban stormwater best management practices: Office of Water EPA–821–R–99–012.

U.S. Environmental Protection Agency, 2000, Bacterial indicator tool: User's guide, EPA–823–B–01–003. U.S. Environmental Protection Agency, Office of Water 4305, 17 p.

U.S. Environmental Protection Agency, 2003, BASINS (Better Assessment Science Integrating Point and Nonpoint sources): accessed March 28, 2009, at http://www.epa.gov/waterscience/basins/.

U.S. Environmental Protection Agency, 2006, Sediment parameter and calibration guidance for HSPF: EPA BASINS Technical Note 8, Office of Water 4305, 43 p.

U.S. Geological Survey, 1994, Pinnacle Mountain topographic quadrangle (Ma): Washington D.C., U.S. Geological Survey.

U.S. Geological Survey, revised 2006, National field manual for the collection of water-quality data: U.S. Geological Survey Techniques of Water-Resources Investigations, book 9, chaps. A1–A9, variously paged. (Also available at http://water.usgs.gov/owq/FieldManual/index html.)

U.S. Geological Survey, 2007, Annual water data reports 2006 water year, accessed September 18, 2009, at http://wdr.water.usgs.gov.

U.S. Geological Survey, 2008, Annual water data reports 2007 water year, accessed September 18, 2009, at http://wdr.water.usgs.gov.

U.S. Geological Survey, 2009a, Annual water data reports 2008 water year, accessed September 18, 2009, at http://wdr.water.usgs.gov.

U.S. Geological Survey, 2009b, National elevation dataset, accessed March 13, 2009, at http://ned.usgs.gov/.

U.S. Geological Survey, 2009c, National hydrogaphy dataset, accessed March 13, 2009, at http://nhd.usgs.gov/.

U.S. Geological Survey, 2010, USGS water data for the Nation, accessed June 12, 2012, at http://waterdata.usgs.gov/nwis.

U.S. Geological Survey, 2012, SPARROW Surface Water-Quality Modeling, U.S. Geological Survey National Water-Quality Assessment (NAWQA) Program, accessed June 26, 2012, at http://water.usgs.gov/nawqa/sparrow/.

Westerfield, P.W., Evans, D.A., and Porter, J.E., 1994, Water resources data, Arkansas, water year 1993: U.S. Geological Survey Water-Data Report AR 93–1, 528 p.

Wilde, F.D., and Radke, D.B., 1998, Field measurements: U.S. Geological Survey Techniques of Water-Resources Investigations, book 9, chap. A6, variously paginated.

Wilde, F.D., Radke, D.B., Gibs, J., Iwatsubo, R.T., 1998a, Preparations for water sampling: U.S. Geological Survey Techniques of Water-Resources Investigations, book 9, chap. A1, variously paginated.

Wilde, F.D., Radke, D.B., Gibs, J., Iwatsubo, R.T., 1998b, Selection of equipment for water sampling: U.S. Geological Survey Techniques of Water-Resources Investigations, book 9, chap. A2, variously paginated.

Wilde, F.D., Radke, D.B., Gibs, J., Iwatsubo, R.T., 1998c, Cleaning of equipment for water sampling: U.S. Geological Survey Techniques of Water-Resources Investigations, book 9, chap. A3, variously paginated.

Wilde, F.D., Radke, D.B., Gibs, J., Iwatsubo, R.T., 1999a, Collection of water samples: U.S. Geological Survey Techniques of Water-Resources Investigations, book 9, chap. A4, variously paginated.

Wilde, F.D., Radke, D.B., Gibs, J., Iwatsubo, R.T., 1999b, Processing of water samples: U.S. Geological Survey Techniques of Water-Resources Investigations, book 9, chap. A5, variously paginated.

Wolynetz, M.S., 1979, Algorithm 139 – Maximum likelihood estimation in a linear model with confined and censored data: Applied Statistics, v. 28, p. 195–206.

www.ingramcontent.com/pod-product-compliance
Lightning Source LLC
Chambersburg PA
CBHW081458170526
45166CB00008B/2474